The theory and practice
of bearings and expansion joints
for bridges

The theory and practice of bearings and expansion joints for bridges

David J. Lee BScTech, DIC, CEng, FICE, FIStructE
Partner, G. Maunsell & Partners

Cement and Concrete Association

Designed and published by the
Cement and Concrete Association, 52 Grosvenor Gardens,
London SW1

Printed in Great Britain by
J. W. Arrowsmith Ltd, Bristol 3

Text set in 10/12 Times New Roman
Headings in 10/12 Times Bold

12.036 First published 1971

SBN 7210 0462 8

Price £2

CONTENTS

FOREWORD

In 1969 the South African Portland Cement Institute in association with the National Institute for Road Research organized a symposium in Johannesburg on the *Engineering of bearings and expansion joints for bridges.* The organizers had become aware, from conversations with bridge and structural engineers throughout South Africa, that there was an urgent need for the dissemination of information on concepts of movement and restraint in bridges and discussion of modern materials and techniques developed in this field.

The author was invited to deliver the principal paper, from which much of the material in this volume is derived.

A lengthy treatise on the subject has not been attempted: rather the aim has been to provide a guide to developments in the field relating modern bridge structures and the current design of bearings and expansion joints. Various references to significant work are given. Similar principles are being used in bridges of various materials. Most bearings and expansion joints are substantially of metal construction and the designer of concrete bridges has therefore to achieve a satisfactory mixed structure of compatible durability and serviceability. Most of the examples refer to concrete bridges for this reason, but the principles are equally applicable to steel, aluminium or timber bridges.

The author is grateful to Dr F. S. Fulton, Director of the Portland Cement Institute, and Dr P. J. Rigden, Director of the NIRR and now Vice-President of the South African Council for Scientific and Industrial Research for allowing the publication of material initially prepared for the Johannesburg Symposium. He would also like to express his thanks to Mr S. Halstead, Director of The Concrete Society of Southern Africa, the staff of the Cement and Concrete Association, and many colleagues in his firm without whose help the volume would never have been written.

INTRODUCTION

Bearings and expansion joints probably cause more difficulties for the bridge engineer than practically any other part of the structure. The service performance of a bridge depends largely on the adequate specification, detailing and careful installation of bearings, drainage and expansion joints.

It is curious that these items are rather more related to the mechanical engineering discipline than to those fields of knowledge which the civil engineer likes to inhabit. Certainly it is true to say that the bridge engineer needs to know more about those aspects of his structures that are giving poor performance. This book attempts an examination of these problems and reviews the application of modern methods.

CHAPTER ONE

Movements

1.1 Brief historical background

Modern requirements for bridge structures have largely generated the problems of bearings and expansion joints: not only are spans larger but such materials as steel and concrete are manufactured by man as monolithic materials. The contrast between these and the fissured and jointed natural materials like stone is striking.

The stone lintel, with its necessarily limited span, laid down shear strength and bending stress as the criteria of design. Movement of the beam under load and temperature change were of no practical significance. When spans were increased with the invention of arch construction they had to be made of relatively small pieces. In consequence, thermal movement took place by a geometrical change of shape in the structure with the small displacements allowed for by dry jointing the stones of the arch and loose filling in the spandrels. When one thinks that such principles have been in use for arch construction since their invention right up to the nineteenth century, it is realized how recent is the problem of the mechanical engineering treatment required for movement.

Timber trusses can also skate round the problem by virtue of the nature of the material. Either the truss can be pin-jointed and the temperature strains taken up with rotation at the joints or, as in some of the medieval hammer-beam roof trusses, complete stiffness can be built into the joints and movements accommodated by flexural distortion.

When concrete began to be used for arches, the brittleness of its monolithic character soon exposed the problem. In Glenfinnan Viaduct, Scotland, the first concrete railway bridge in Britain, 21 arches of 15 m (50 ft) span are formed of mass concrete. The possibility of differential settlement of the piers was dealt with by incorporating a sliding joint at the crown of each arch consisting of two 13 mm ($\frac{1}{2}$ in.) thick steel plates placed together.(1) Since the viaduct was opened in 1898 there has been no noticeable movement of the piers and the horizontally curving structure is still in good condition.

Structural engineering in the past had to grapple with the stability of foundations to a far greater extent than problems of accommodating thermal strains in the structure once it was standing. History then does not provide much background to what is essentially a modern problem.

With highway bridges, not only has the intensity of loading tended to increase but so also have the velocity and frequency of traffic, to such an extent that fatigue and dynamic loading are beginning to be subjects of concern and will become increasingly so as the strength/weight ratio of steel or concrete is improved.

All these factors mean that modern bridges are more sensitive to the movements that arise in them, and the generation and magnitude of these movements are of increasing importance.

1.2 Sources of movement in structures

An essential first step in assessing the problems of movement in structures is to set out the sources from which movements are likely to arise in any given situation and to set down quantitative data which will fulfil the requirements.

One might think first of the distortions caused in a structure related to the elastic and plastic properties of the materials used, such as:

- thermal and humidity changes;
- creep, shrinkage and fatigue;
- axial and flexural strains arising from dead and live loading including the effects of prestressing;
- dynamic effects including impact, braking and lurching, centrifugal forces, etc.;
- provision for overload.

Next one may set down the exterior sources of movement, such as:

- tilt, settlement or movement of ground;
- mining subsidence;
- seismic disturbances;
- structures designed with moving parts, i.e. lift spans and swing bridges;
- erection procedures.

The question of whether movements are related to long-term or short-term effects requires careful consideration.

The directions of movement, whether they are reversible or not, and what rotations are associated with components of movement must all be determined.

For the design of a prestressed concrete box beam spanning approximately 60 m (200 ft) in an area of mining subsidence, it is of interest to record the movements in the longitudinal direction which were taken into account.

(a) Temperature ± 13 mm ($\pm\frac{1}{2}$ in.).
(b) Humidity ± 3 mm ($\pm\frac{1}{8}$ in.).
(c) Residual creep and shrinkage -25 mm (-1 in.)
(d) Elastic prestressing deformation of concrete -13 mm ($-\frac{1}{2}$ in.)
(e) 0·2% ground strain owing to mining subsidence ± 150 mm (± 6 in.).

These have been quoted for a particular instance to show how the various figures can add up. Thermal changes become progressively more important for large spans and continuous bridges. They are dealt with in some detail in this volume as recent investigations have given valuable data on which to base design criteria for thermal movement.

Modern codes of practice appear to be reasonably adequate in estimating movements of concrete under stress, as the economic use of prestressing has generated an interest in the calculation of losses arising from change of strain.

For special concrete structures, such as large arches like Gladesville Bridge,(2) studies of the behaviour of the structure under load has led to a valuable extension of knowledge.

1.3 Temperature movements

In general, it is not practical to restrain a structure fully against temperature movements. This becomes obvious if we consider a prismatic concrete beam of cross-sectional area 0·93 m^2 (10 ft^2). If this beam heats up by 10 degC (18 degF) and is restrained from expanding axially, a stress of 4·34 N/mm^2 (630 lbf/in^2) is induced and the restraining force required is 4·04 MN (405 tonf). These figures are based on a coefficient of linear expansion of $12·6 \times 10^{-6}$ per degC (7×10^{-6} per degF) and a Youngs modulus of 34,500 N/mm^2 (5×10^6 lbf/in^2).

However, in practice, some restraint is inevitable since completely frictionless bearings do not exist. In order to estimate the effect of these restraints, it is necessary to know what movements would take place if the structure were totally unrestrained. The actual movements of the structure can then be found by adding the movements caused by the restraining forces. Methods of estimating the restraining forces will be discussed later.

It is well known that if a monolithic body heats up evenly, each part of the body will move along a line radiating from the point of zero movement. Also, the movement is proportional to the distance from the point of zero movement and to the temperature change. Hence, if we know the geometry of the body, the coefficient of linear expansion and the temperature change, it is easy to calculate the movement.

Another effect is thermal bending, which can occur when different parts of a structure are at different temperatures. The temperature distribution across a cross-section will generally not be linear and hence an internal system of stresses will be induced such that the variation of strain across the section is linear. This will, of course, only apply if we accept the assumption that plane sections remain plane, which is a usual design assumption. Although the induced stresses are of importance in the design of the section, as far as the movements produced are concerned, the only effect is that the structure takes up a vertical curve of constant radius R where:

$$R = \frac{D}{\Delta_\varepsilon}$$

where

D = overall depth of section

Δ_ε = difference in strain between extreme fibres

It can be seen from this that in order to calculate the 'free' thermal movements we must know:

the coefficient of linear expansion of the material;
the range of mean structure temperature that can occur;
the worst possible temperature distribution.

In Britain we are guided in these matters by BS 153: Part 3A: 1954(3) and by a Ministry of Transport Memorandum.(4) The relevant clauses from these standards are reproduced in Appendix 1 for comparison with the standards in use in Germany and the USA.

Over the last few years, the Ministry of Transport's Road Research Laboratory has been, and still is, conducting a programme of research into bridge temperatures and movements. This has included the measurement over a period of years of movements on a number of major structures throughout Britain. The results of this work so far are contained in various reports, some published by the Ministry of Transport.(5–9)

One striking feature of this work is that it has shown our standards to be inadequate in their recommendations. In particular, the minimum temperatures can be lower than specified and the temperature difference between parts of a section can be higher, although other concrete properties such as shrinkage and humidity changes somewhat mitigate the effects of extreme temperatures in the United Kingdom.

For steel structures, a coefficient of linear expansion of $11·7 \times 10^{-6}$ per degC ($6·5 \times 10^{-6}$ per degF) is confirmed by the measurements to be adequate.

For concrete structures the coefficient of linear expansion has been found to vary between 13×10^{-6} per degC ($7·2 \times 10^{-6}$ per degF) and 8×10^{-6} per degC ($4·4 \times 10^{-6}$ per degF) for the Hammersmith Flyover in London (gravel aggregate) and for the Mancunian Way in Manchester (limestone aggregate) respectively. This confirms work carried out by the Building Research Station.(10)

The research programme has also developed an iterative method for predicting the temperature distribution through sections. This requires a knowledge of heat-transfer coefficients, intensity of solar radiation and shade temperatures. The values of heat-transfer coefficients have to be estimated for the particular materials being used, values for both concrete and steel surfaces being quoted in references 5 and 6 which, when used for particular bridges, gave good agreement with measured temperatures. The intensity of solar radiation is dependent on the latitude in which the structure is situated, its orientation, the season of the year and the hour of the day. The shade temperature is also dependent on the location of the structure and, as with solar radiation, reference must be made to local meteorological records.

Once these factors are known, the temperature distribution and hence the mean temperature can be determined for any cross-section. Fortunately, however, the reports come to some simplified conclusions regarding the range of mean bridge temperatures. These are as follows.

(1) For concrete bridges, both the experimental and theoretical results suggest that the range of extreme mean bridge temperature is much the same as the range of extreme shade temperature in the same area.

(2) For steel box-section bridges, both the experimental and theoretical results suggest that:

on cold winter days, the minimum mean bridge temperature will fall to 3 to 4 degC below the minimum shade temperature;

on hot summer days, the maximum mean bridge temperature will be about 1·5 times the corresponding maximum shade temperature in degrees Celsius.

(3) The ranges of bridge temperatures quoted in the current British Standards have been shown to be inadequate. In particular, the minimum temperatures are not low enough and the ranges quoted are not appropriate to any area of the British Isles.

(4) For the purpose of setting joints and bearings, the times when the mean bridge temperature is approximately equal to the shade temperature are:

for a concrete bridge: 0900 BST ± 1 hour each day;

for a steel bridge: at about 0400 to 0600 BST each day during the summer and at any time on 'average' days during the winter (i.e. excluding very cold nights and warm days).

Hence, the maximum range of mean bridge temperature can be estimated from a knowledge of the maximum and minimum shade temperatures likely to occur in the region of the structure. In Britain we are particularly fortunate that records of maximum and minimum shade temperatures are available over a period of more than a century.

There is, however, no simple way to determine the temperature distribution. For solid sections, an exponential distribution has been suggested[11] which gives a reasonably simple result.

For flanged sections we can make the simplifying assumption that the top and bottom flanges are at constant mean temperatures differing by ΔT. This temperature distribution will give approximately the same effects as the true distribution, if the ratio of flange thickness to overall depth is small, with the exception that it underestimates the induced compressive stress in the extreme top fibre.

The value of ΔT will depend on the section properties but values of 10 degC have been observed in structures in Britain, indicating that the recommendations of BS153: Part 3A: 1954 might be revised in this respect.

1.4 Restraints associated with thermal bending

It has been shown above that, owing to differential temperature effects, a structure will want to take up a vertical curve of radius R. If the structure is statically determinate, as in a simply supported beam or a cantilever, this can happen freely, producing end rotations.

In this context it is appropriate to point out that cantilevers are particularly sensitive to this sort of deflection. The leading cantilever of a continuous 61 m (200 ft) span concrete viaduct in London was observed to deflect by 13 mm ($\frac{1}{2}$ in.) during the day and required a propping force of approximately 1 MN (100 tonf) to prevent this movement. It is also of interest to note that, during the construction of the Beachley Viaduct (a steel box-beam deck), a temporarily cantilevered portion of deck was reported to oscillate perceptibly as clouds moved across the sun.

If, however, the structure is statically indeterminate, as in a continuous bridge, the redundancies will cause restraint.

In bridge decks this problem can be solved by distributing the fixed end moments for each span. The fixed end moment M is independent of span length and is given by the formula

$$M = \frac{EI}{R}$$

The distributed moments produce rotations and deflections which can be calculated by means of standard methods. They also produce changes in the support reactions, which can be of significance. In order to give some idea of the magnitude of these effects, an example of a simple two-span concrete box-beam bridge is given below. In this case, it can be seen that a temperature difference of 10 degC between top and bottom slabs produces support reactions of as much as 30% of the self-weight reactions.

1.5 Example showing effect of thermal bending

Consider the following two-span structure:

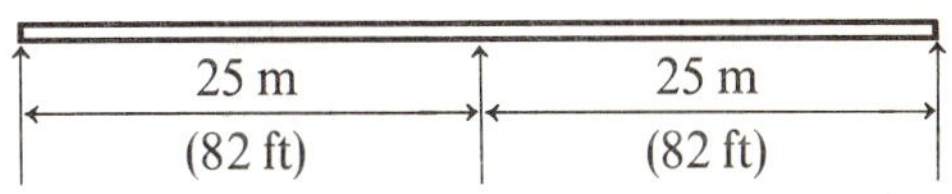

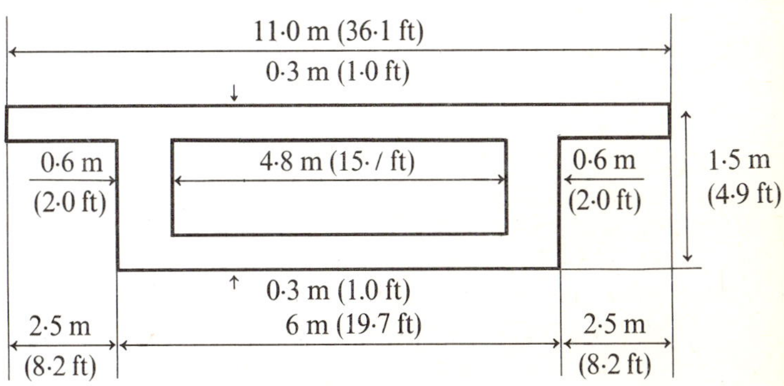

Section properties

Area	$= 6{\cdot}18\ m^2$ ($66{\cdot}5\ ft^2$)
Height of neutral axis	$= 0{\cdot}90$ m (3·0 ft)
Moment of inertia	$= 1{\cdot}82\ m^4$ ($210\ ft^4$)
Self-weight	= 14,850 kg/m (10,000 lb/ft)
Coefficient of expansion	$= 12 \times 10^{-6}$ per degC ($6{\cdot}7 \times 10^{-6}$ per degF)
Young's modulus	$= 34{,}500\ N/mm^2$ ($5 \times 10^6\ lbf/in^2$)

Assume mean temperature of top flange is 10 degC (18 degF) higher than the rest of the section. Hence, strain differential between extreme fibres

$$\Delta_\varepsilon = \frac{1{\cdot}5 \times 10 \times 12 \times 10^{-6} \times 3{\cdot}3 \times 0{\cdot}45}{1{\cdot}82}$$

$$= 150 \times 10^{-6}$$

Radius of curvature

$$R = \frac{1{\cdot}5}{150 \times 10^{-6}} = 10{,}000 \text{ m } (32{,}800 \text{ ft})$$

Fixed end moments

$$M = \frac{34{,}500 \times 1{\cdot}82}{10{,}000}$$

$$= 6{\cdot}26 \text{ N mm } (4{\cdot}6 \times 10^6 \text{ lbf ft})$$

Moment at central support $= 9{\cdot}39$ N mm ($6{\cdot}9 \times 10^6$ lbf ft)

Changes in support reaction are:

central support	−750 kN (75 tonf)
end support	+375 kN (38 tonf)

Self-weight support reactions are:

central support	4500 kN (456 tonf)
end support	1365 kN (137 tonf)

Hence changes as a percentage of self-weight reactions are:

central support	−16%
end support	+27%

End rotation = 0·000625 radians.

1.6 Erection methods

A brief mention might be made of some of the methods whereby controlled movements may be introduced in a structure during construction for obtaining the required distribution of strains or stresses. After completion, the replacement of defective elements or access for maintenance and inspection purposes may be required.

Hydraulic jacks

In many bridges, provision has to be made for jacking in order to modify stresses or strains in the structure or to counteract subsidence. Similarly, the newer types of bearing generally bring with them the prudent requirement that provision has to be made for replacing them should they give trouble during the life of the bridge. Hydraulic jacking is the usual method by which these movements are achieved. Hydraulic jacks may vary in capacity from very nominal values up to well over 1000 tonnes or more. In all cases, care must be taken in the design of the structure to allow space for inserting a conventional jack; also, the reactions of the jack must be properly taken into account through the deck and substructure. The designer should not leave himself in the position where a burst or leaking hydraulic jack presents an unforeseen hazard. Locking, propping and wedging should be employed to prevent jamming of the jack through any failure. Similarly, it should be remembered that there will be a loss of pressure when the system stands for any appreciable period of time.

Flat jacks

These devices derive from the brilliantly inventive engineer Eugène Freyssinet. Full information about them is not given in this volume as it may be easily obtained from the manufacturers. A notable application in bridge construction was the jacking of the arch ribs of Gladesville bridge in order to strike the centring.(2) Flat jacks are inexpensive, but it is even more important to provide a dual system so that, if one circuit fails, the load can be safely taken by the duplicate system.

Sand jacks

These cheap devices have been found very satisfactory when controlled lowering is required, such as lowering beams onto their bearings or striking falsework. Provided the sand is maintained in a dry condition, their performance is extremely reliable and it is not even necessary to provide a duplicate system. In the British climate, the sand is kept dry simply by covering the body of the jack with a polythene cover. They were extensively used on the Western Avenue Extension project.(6)

above: One of Brunel's timber viaducts, Penryn, Cornwall.
below: Glenfinnan Viaduct, Inverness-shire, Scotland

CHAPTER TWO

Flexibility and articulation of the structure

2.1 Structural flexibility

Bridge structures of welded high-tensile steel and prestressed concrete must be designed to have sufficient flexibility to permit temperature strains and certain live-load strains to occur with the minimum of restraint. This is particularly true of the more complex torsionally stiff structural forms that are becoming more important. Neglect of these strains in modern bridges may lead to expensive maintenance problems or, more seriously, to failure by brittle fracture due to biaxial or triaxial tension. It is a new problem, because riveted mild-steel and reinforced concrete bridges are not only more flexible and ductile than their more modern counterparts but are of relatively unsophisticated configuration.

A modern example of internal flexibility is the corrugated web plate which is stressed in shear by dead and live loads and not affected even by severe differential temperature strains.

An example of external flexibility is the system of vertical links supporting the stiffening girder of the Severn suspension bridge. These permit longitudinal movements to take place freely at both ends thus reducing the lattice truss action of the inclined hangers. This prevents overstressing of the hangers by live loading.

The use of new materials such as fibre-reinforced concrete with a high tensile strength but low ductility may lead to more brittle structures. The possibility of a sudden collapse of such a structure can be removed by designing flexibility into the structure in the form of a 'fuse plug' member of ductile material to act as an energy-absorbing hinge. In a less dramatic sense, this concept is being developed by limit-state methods to achieve greater economy for box-girder bridges of conventional materials.

2.2 Arrangement of structures to take account of movements

The first matter to consider is whether a structure may be arranged without joints and bearings so that thermal changes can be accommodated by distortion and restraint of the structure. With concrete, the reversibility of thermal effects suggests that the imposed stresses do not get time completely to disperse by creep. Certainly it is conventionally assumed that the restraining stresses cut into too big a proportion of the working stress of the structure to allow such a simple solution to the problem. Probably the commonest engineering use of a fully restrained structural element is in long welded rail track.

Arch construction has been mentioned as one historical answer where the structure is capable of distortion under temperature change without imposing excessive stresses in the structural elements or force on the restraining abutments. Similar ideas may be applied in tall, spindly columns and certain types of portal frame provided there is a sufficient factor against buckling.

The historical reference to very short-span structures requiring no joints or joints of very limited travel can be applicable. In Britain, for instance, bridge spans up to 9 or 12 m (30 or 40 ft) require no specific allowance for bearings. With precautions for rotation, highway surfacing can be carried over the joints between beam and abutment for spans of up to 18 m (60 ft). In the northerly parts of the United Kingdom, the span can be increased to 30 m (100 ft). Clearly, climatic conditions and form of construction can weigh very heavily in the engineering of short spans.

At the other extreme, the advantages of building very long strings of spans with full continuity poses the problem of designing bearings and expansion joints with a large capability for movement. The bearings and expansion joints for this type of structure have received considerable attention in recent years.

2.3 Short columns

The assumption of fixity at the top or bottom of a short column built in either to the deck or to the foundation, can be shown to be misleading. For columns built into the deck the correct flexibility ratios would develop from frame analysis but the foot of the column should similarly receive realistic treatment.

Leonhardt[12] has shown that it is possible greatly to relieve the fixed-end moment for spread footings by taking into account the modulus of reaction of the soil as a spring loading. Leonhardt suggests that with high modulus of reaction and a narrow foundation the assumption of a pin joint at the base is more correct than the assumption of fixity. In most instances, some degree of fixity is present and provided that the modulus of reaction of the soil is available there is no difficulty in analysis. If, however, soil characteristics are not available, it is misleading to take the limiting encastré case and assume this is on the conservative side.

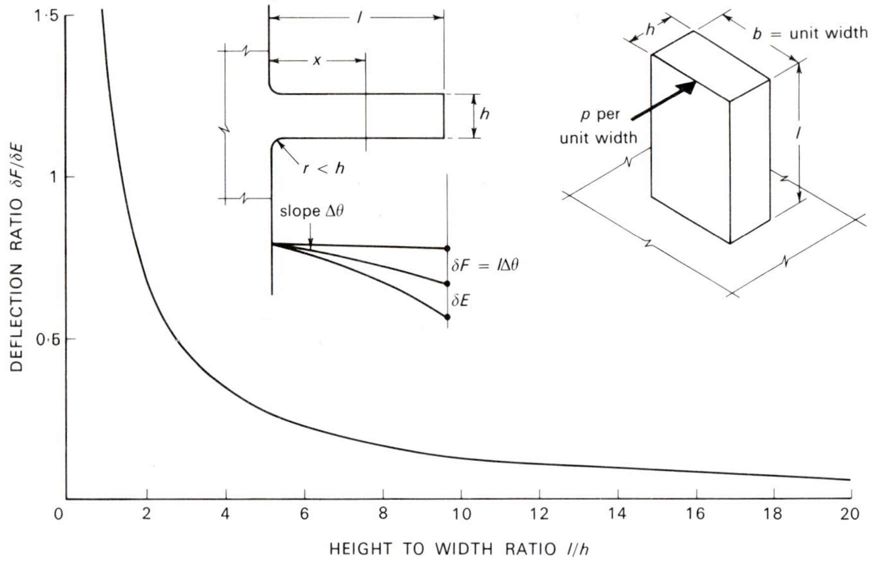

Figure 1
Relation between deflection ratio and height/width ratio.

If the column is of concrete, it is reasonable to assume a supporting medium of similar elasticity to the material of the column. For short columns continuous with a spread footing or pile cap, it is quite simple to take into account this assumption of elasticity of the support.

Consider that the base of the column is built into a semi-infinite supporting material of the same type as the column (Figure 1).

When fixed-ended, the deflection at the top $Pl^3/3EI = 12Pl^3/3Eh^3$.

If we put $l/h = \gamma$, then deflection $\delta_E = 4P\gamma^3/E$.

There is an additional deflection caused by the elasticity of the supporting medium.

Roark[13,14] gives

$$\Delta\theta = \frac{16{\cdot}67Pl}{\pi E h_1^2} + \frac{(1-\nu)P}{Eh_1}$$

where $h_1 = h + 1{\cdot}5r$. In this example, the root fillet radius $r = 0$. This additional deflection is $l\Delta\theta$ at the top. Hence, deflection

$$\delta_F = \frac{16{\cdot}67Pl^2}{\pi E h^2} + \frac{(1-\nu)Pl}{Eh}$$

$$= \left(\frac{16{\cdot}67\gamma^2}{\pi} + (1-\nu)\gamma\right) P/E$$

The increased deflection at the top is thus in non-dimensional terms

$$1 + \frac{\delta_F}{\delta_E}$$

where

$$\frac{\delta_F}{\delta_E} = \frac{\left(\dfrac{16{\cdot}67\gamma}{\pi} + (1-\nu)\right)}{4\gamma^2}$$

For concrete, Poisson's ratio ν is approximately 0·15. Hence,

$$\frac{\delta_F}{\delta_E} = \frac{5{\cdot}3\gamma + 0{\cdot}85}{4\gamma^2}$$

This relationship is illustrated in Figure 1.

It will be seen that, for the short columns often used for elevated roads, the increase in deflection when elasticity at the foot is taken into account can alone be anything between 25 and 75 % greater than the classical bending deflection assuming full fixity. This in itself reduces the required travel of bearings and the forces generated. Pile caps and pile foundations would generally be more flexible than the assumption of an infinite elastic medium and this could be assessed separately by investigating the frame action of the pile group.

It will be clear, however, that the actual rotation generated at the top of a column is greater than when base fixity is assumed.

For the precise assessment of the behaviour of short columns, the elastic foundation can be properly dealt with by the insertion of spring supports for tilt, vertical and lateral translations such as given in standard texts (see p. 118 of reference 15).

Where forces are not reversible, the expressions for deflection after creep have also been given (see p. 877 of reference 15).

2.4 Tall slender columns

In the Tasman Bridge, Hobart, the columns of the approach spans are from 12 to 38 m (40 to 124 ft) high springing from a pile cap at water level and with a steel rocker bearing absorbing rotation changes from the continuous prestressed concrete deck. The columns are high enough for the assumption of full fixity at the base to be perfectly valid. The bending deflection of the concrete columns provides the movement capability. The slenderness ratio is up to 50, well outside the normal code of practice provisions for columns.

However, by this arrangement of the articulation, the column is made determinate and can be designed for the actual moments obtained. The design may be based on either the ultimate strength of the section or the yield stress of the material, with a suitable load factor. The

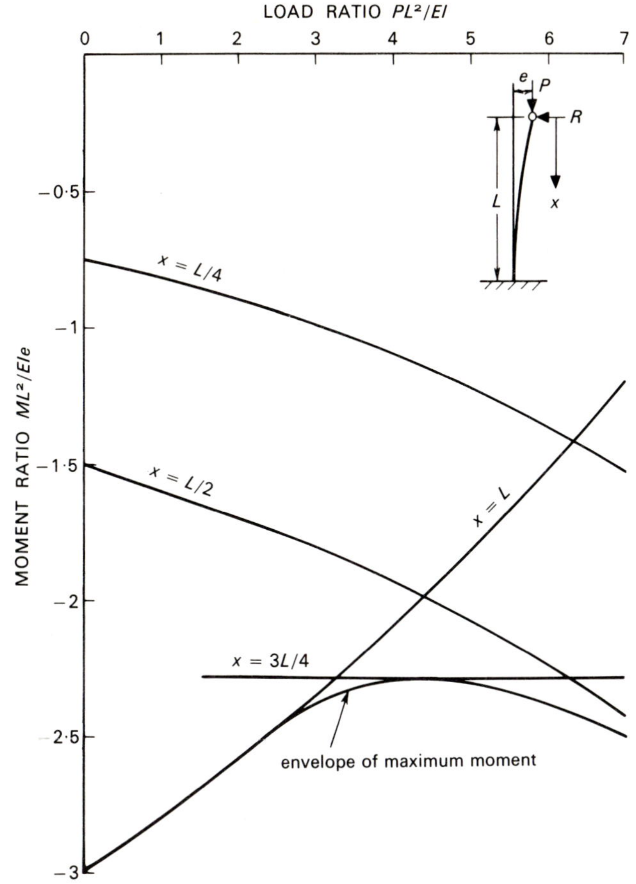

Figure 2
Design curves for deflected column under normal load.

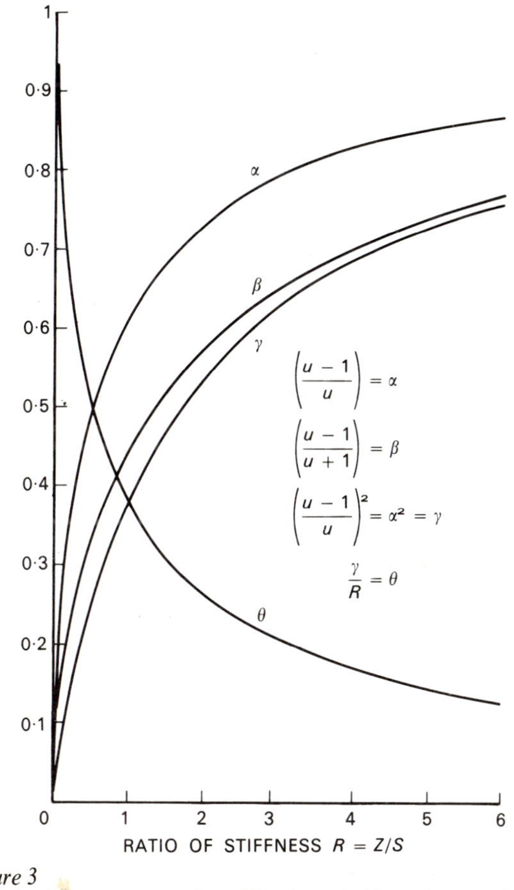

Figure 3
Relation between proportion of load carried by column and stiffness ratio

analysis was based on normal elastic theory, resulting in the design curves shown in Figure 2 which are self-explanatory.

The analysis for the bridge column design also took account of lack of straightness in construction and accidental eccentricity of prestress. Vertical prestress was provided to ensure that the concrete columns exhibit proper elastic behaviour.

2.5 Piers with twin flexible walls

A slim vertical pier poses the problem of flexibility for taking up movements but having sufficient elastic stability. Muller[16] has shown how an optimum relationship of these factors can be achieved by the geometrical layout, with particular reference to the use of twin inclined piers in the bridge over the Seine at Choisy-le-Roi.

2.6 Viaducts with flexible columns and bearings

The distribution of traction forces along a flexible viaduct can be simply derived by means of the graphs in Figure 3: these are based on the assumption that the viaduct is long enough for effects at either end to be independent; the deck members are mounted on flexible bearings at one end of each span and fixed bearings at the other.

If we apply a traction force P to the span connected to some column X near the centre of the viaduct, then the load carried by this column is βP. The load carried by columns adjacent to X falls off in a geometric progression of the form

$$\frac{\beta P}{u}, \frac{\beta P}{u^2} \cdots \frac{\beta P}{u^n}$$

where

$$u = \left(1 + \frac{R}{2}\right) \pm \sqrt{R + \left(\frac{R}{2}\right)^2}$$

The value of β is plotted against R in Figure 3, where $R = Z/S$, the ratio of the column stiffness (Z) to the sum of the shear stiffness (S) of all the bearings under the loaded span.

If an abutment of a long viaduct is rigid, and the end span is supported on it on flexible bearings, the load carried by the end column when a traction force P is applied to the end span is γP, and the force transmitted to the abutment is θP.

If the abutment is flexible (the same stiffness as the columns) and a traction force P is applied to it, the load carried by it is αP, and the load carried by the end columns is $\alpha P/\mu$.

Zederbaum[17] has also considered the elastic distribution of forces and displacements. This approach is essential to a thorough assessment of structures incorporating elastic bearings.

2.7 Horizontal forces on continuous beams supported on sliding bearings

In the Mancunian Way elevated structure,[18] which has a compositely curved horizontal alignment, an investigation was made into the forces acting on the bearings on the columns during movements of the structure. A series of computer analyses were made to establish the effect of flexibility in the substructure, omission of restraints at intermediate columns, variations in friction coefficient, clearances in the bearings, and optimum alignment of the bearings. A plane frame program treated the structure as polygonal and replaced the tapering columns by spring members with stiffnesses appropriate to their height. It was found convenient to consider the resulting forces under two headings:

the expansion and contraction forces required to constrain the structure to move through the expansion joint at the free abutment if bearing friction is ignored;
the forces resulting from bearing friction.

Expansion and contraction forces

These forces would be zero if the structure were free to expand in any direction on frictionless bearings. In practice, lateral restraints are required to carry wind and centrifugal forces, even if not necessary to control the alignment of the structure at the expansion joint. Initially the structure moves freely to take up the lateral clearances in the bearings. Further movement produces an equilibrium set of restraint forces on the bearings. As can be seen from the typical restraint force curve in Figure 4, the sign of the forces changes several times. Thus, the bearing clearances on opposite faces must be closed before a balanced set of forces can develop and this must be borne in mind when combining restraint forces with the wind and centrifugal forces which are unidirectional.

The investigation proved that an increase in the bearing clearances or a reduction in the substructure stiffness causes a reduction in the restraint forces. It was also shown that, in theory, the forces can be minimized by orienting the bearings on the column caps to lines which are not necessarily tangential to the structure. However, for a multi-curved structure, the optimum line is not normally worth the trouble of setting out.

Friction forces

In a long straight structure, the thrust which accumulates from the forces due to bearing friction causes no lateral load on the bearings. In a curving structure, however, it was shown that significant transverse forces arise which depend on the angular deviation of the structure (see Figure 5). The friction forces, which increase towards the fixed end, do not form a balanced set and are not influenced, apart from local redistributions, by variations in bearing clearance or substructure stiffness.

Figure 4
Distribution of lateral forces on columns due to movement of deck.

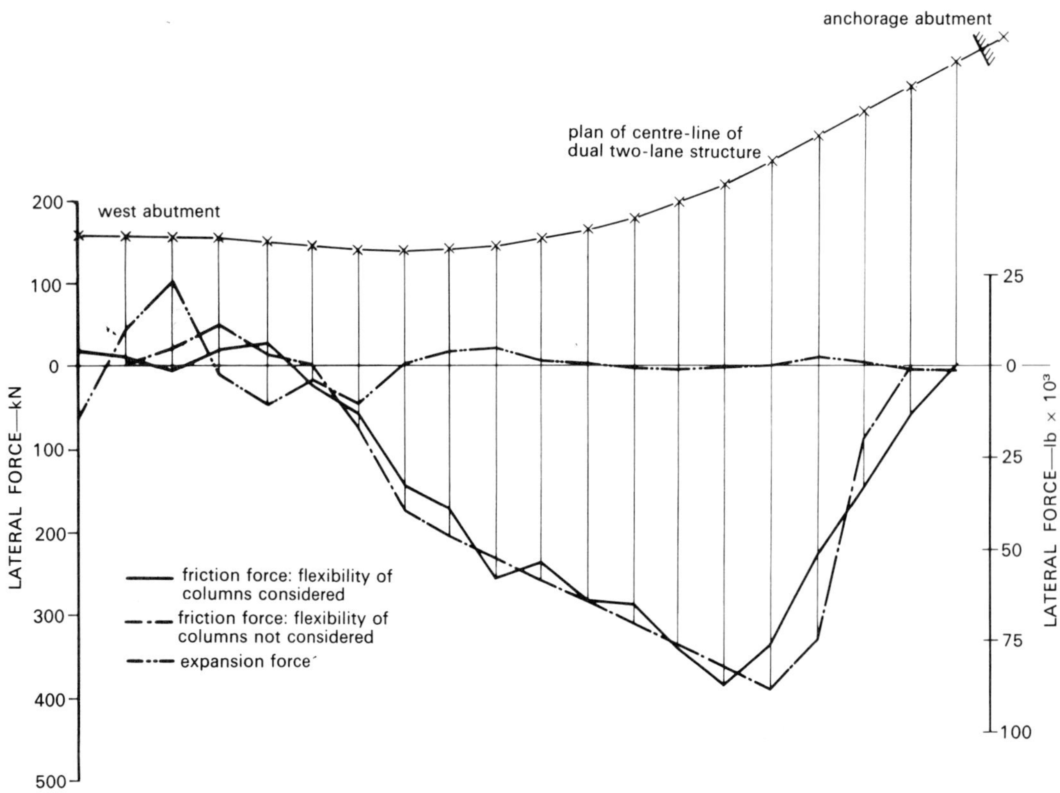

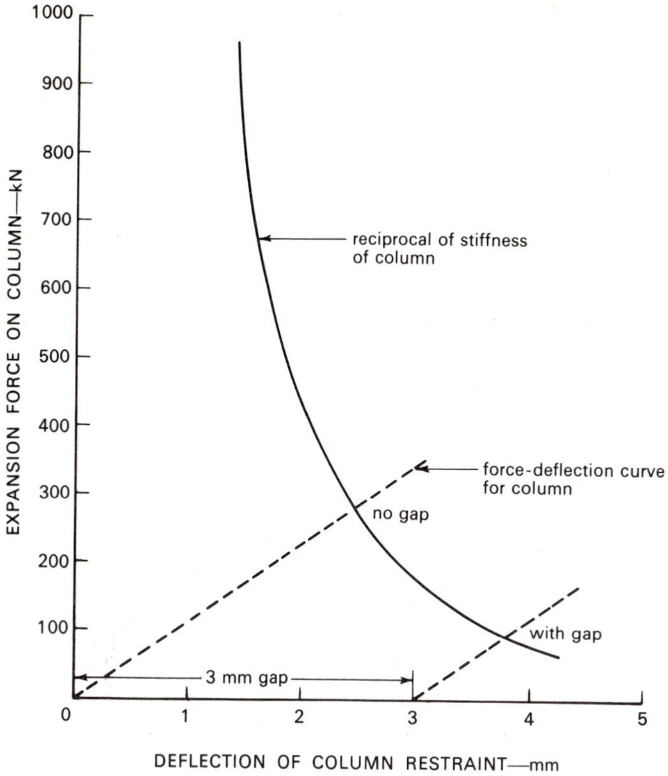

Figure 5
Effect of a gap on the expansion force of a typical column of the Mancunian Way.

right: Tasman Bridge. View of elastically flexible column.
below: Tasman Bridge, Hobart, Tasmania.

above: Choisy-le-Roi Bridge under construction over the Seine near Paris.

left: Mancunian Way elevated road, Manchester, England. View during construction showing Rotaflon bearings in position at top of columns.

CHAPTER THREE

Expansion joints

3.1 General principles

The expansion joint must be recognized as an integral part of any bridge structure and as such should be considered at an early stage in the design. If the expansion joint is carefully designed and detailed, properly installed by specialist operatives and given reasonable maintenance in service, there is no reason why it should not give trouble-free performance for many years.

However, in practice the expansion joint is often casually chosen at the last moment, badly installed by unskilled labour and neglected in service. Inevitably, with the passage of time it gives trouble and the tendency is to condemn that particular type of joint as unsatisfactory. This is a common situation not unique to any particular form of joint and experienced in most countries. The reason why remedial measures are more successful is often that more care is taken in implementing corrective works than was taken with the original joint.

It is important to appreciate that the expansion joint is in the most vulnerable position possible on any bridge, situated at surface level where it is subjected to the unabsorbed impact and vibration of the traffic and exposed not only to dust, silt, grit and water but also to the effects of ultra-violet rays, ozone attack and chemicals such as salt solutions, cement alkalis and petroleum derivatives.

In assessing this situation, we derive the following requirements for an expansion joint. It must:

accommodate all movements of structure and withstand all loadings;
not impart stress to the structure unless the structure has been designed accordingly;
have good riding characteristics;
not present a skid hazard;
not present a danger to traffic such as cyclists, animals, etc.;
be silent and vibration-free in operation;
give reliable operation throughout the expected temperature range;
be sealed against water and foreign matter or make provision for their disposal;
resist corrosion and withstand attack from grit, chemicals, etc.;
facilitate easy inspection, maintenance and repair.

The Road Research Laboratory has gained valuable knowledge from studies of the performance and design of expansion joints in conjunction with the experience of various manufacturers and engineers in Britain and abroad. At the present time our policy on expansion joints is largely guided by a document[4] issued by the Ministry of Transport in May 1967 which, largely based on field studies at that time,[19] sets out recommendations for both the design and construction of such joints.

Today there is a large variety of proprietary expansion joints on the market and the problem facing the engineer is now so often that of selecting the most suitable joint to give good performance and a trouble-free life for at least as long as that of the surfacing.

Köster[20] catalogues many joints in detail and Watson[21,22] provides a valuable commentary in recording the disastrous effects of failure in joint details. In the light of this information on particular joints, this volume deals with the general approach to the jointing problem rather than comments on proprietary products.

3.2 Problems of geometrical layout

The expansion joint has to accommodate movements of the structure in the longitudinal, transverse, vertical and rotational modes. Analytical methods of assessing the displacements and forces associated with this structural behaviour have already been discussed, but there are several features, mainly concerning the geometry of the structure and the joint, which require careful consideration if trouble-free operation is to be achieved.

With structures curved in plan or with skew joints, the relative movement may not be normal to the line of the joints (Figure 6a and b). This can lead to binding of the elements or high shear forces in filler materials which may be exuded and carried off by the traffic, so causing ultimate failure. It is therefore important to assess this transverse displacement and to design the joint accordingly or eliminate the movement by restraining the bridge either at the joint or preferably at the bearings. A similar type of problem may be experienced through excessive flexural rotation at the joint (Figure 6c) but, by arranging the end bearing and expansion joint in the same vertical plane, the discontinuity in the vertical direction can be minimized and the motion reduced to a purely horizontal displacement. On joints designed for small movements, this may also be overcome by providing an articulated running-on slab.

As it is usual to set the bearings horizontally a discontinuity may be introduced at the expansion joint if

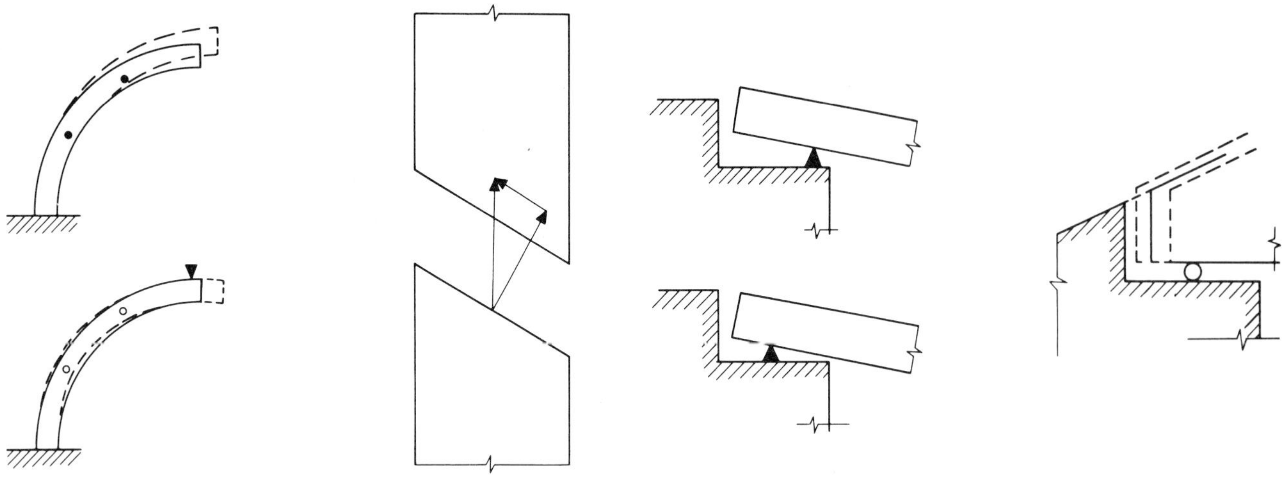

(*a*) (*b*) (*c*) (*d*)

Figure 6
(a) Unrestrained and restrained movement of a curved structure.
(b) Normal and shear displacements across skew joint affecting filler material.
(c) Position of bearing affects discontinuity of joint due to end rotation.
(d) Surface discontinuity at joint on inclined carriageway.

Figure 7
(a) Detail of joint suitable for small movements.

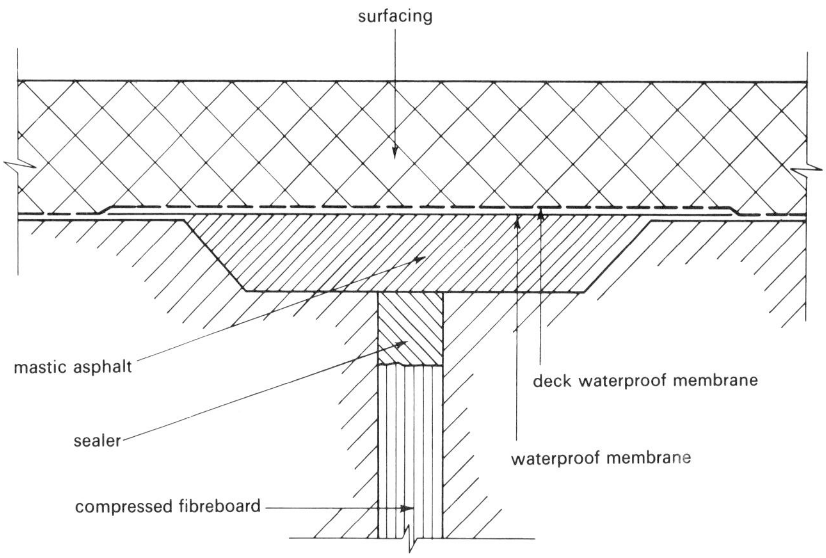

(b) Cantilever joint set back to reduce edge stresses.

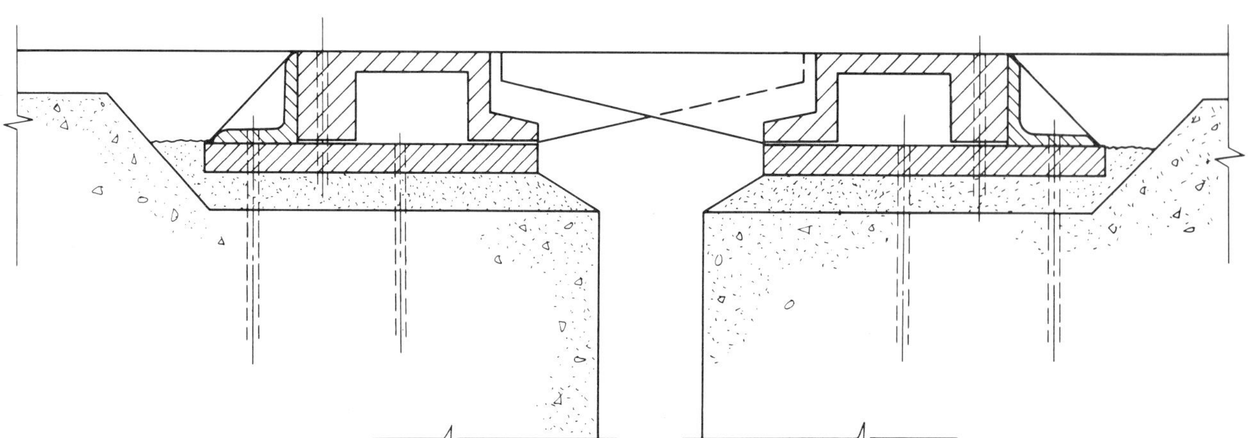

the structure is on a vertical grade θ (Figure 6d). This problem is significant if the total movement M is large because, with the joint set at grade for the mean position, the magnitude of the vertical discontinuity is given by $(M \tan \theta)/2$. There may be a sudden jump or a change of grade over a certain length depending on the type of joint. The problem can be overcome by sloping the bearings if the supporting structure can withstand a lateral force applied in the same direction as the upward slope. Alternatively, a compensating mechanism such as a wedging device may eliminate or reduce the effect to an acceptable magnitude but this usually means added complication and expense.

3.3 Joints for small movements

For movements of less than 5 mm (0·2 in.), it is usually considered that no special provision is necessary, and for movements up to about 10 mm (0·4 in.) the most popular treatment is for a gap-filled joint with continuous surfacing. This can be a very satisfactory joint if carefully formed; the filling is protected by the surfacing which also absorbs much of the impact. The usual practice is to fill the gap with compressed fibreboard with a poured sealant closing the gap at the top. The sealant must be applied with care to properly prepared surfaces. As water usually percolates through the road surfacing, the joint should be made wherever possible to a natural fall to assist drainage. The joint itself will not be waterproof, so it is always expedient to provide drainage under the joint to avoid staining on abutments and columns.

On some joints of this type, it has been fashionable to provide a shallow sawcut in the top of the road surfacing which is filled with a sealant. This provides a plane of weakness to confine cracking but in some cases the sealant has been pulled out by suction from vehicle tyres and has initiated breakdown of the surfacing. In all fairness, however, several of these joints have given years of excellent service. It is for conjecture that modern low-pressure radial-ply tyres are more damaging to the surface sealant than the more conventional tyre. There seems no objection to omitting the sawcut, thus inducing micro-cracking over the debonded length of surfacing. On many of our bridges in the UK we employ a continuous waterproof membrane over the deck area, largely to prevent the concrete from attack by saline de-icing chemicals. Particular care is required to maintain continuity of the membrane over the joint. The waterproof layer is usually a plastic, bituminous felt or copper membrane, liquid treatments generally being confined to conditions less rigorous than bridge decks. A detail of this type is shown in Figure 7a where recesses are left in the mating units and filled with mastic asphalt to ensure a smooth profile across the joint.

3.4 Joints for medium movements

The most popular type of joint in this range is undoubtedly the preformed sealer, compressed or clamped between nosings. There is a large range from which to choose and some can be joined back to back to give an increased range of movement. These joints are particularly suitable where pedestrian, cycle or animal traffic has to be accommodated as they provide a continuous surface. However, the engineer should satisfy himself that arrangements for accommodating kerbs, edge beams and medians are adequate as it is often there that trouble will start.

Most of the cellular fillers are based on rubber or neoprene and, although the normal wearing properties are good, they are subjected to severe treatment and prone to damage; provision for easy maintenance is therefore essential. It is the author's opinion that these joints should be conservatively regarded as gritproof rather than waterproof. The compressed member cannot be regarded as waterproof, and only where the member is continuous throughout its length and mechanically clamped will it be an effective water barrier. It is therefore prudent to provide at least elementary drainage on the underside and to arrange surface slopes and gully positions so as to prevent as much water as possible from reaching the joint.

These joints require planned routine inspection if they are to perform satisfactorily, and any large deposit of silt and grit, particularly around kerb details, should be removed, because it can severely restrict the free movement and may support vegetation growth which can damage the sealer.

In using precompressed sections the forces exerted on the structure by the sealer should not be ignored; the modulus of poured sealants and the hollow tube elastomers is usually of the order of 170 N per mm compression per m length. For example, a sealer compressed 25 mm (1 in.) over a 15 m (50 ft) wide bridge could exert a force of some 64 kN ($6\frac{1}{2}$ tonf) at surface level of the structure. Manufacturers of such products often do not quote the modulus of their products in their sales literature.

Since this is a surface-level joint, the fixing arrangement requires careful detailing and skilled installation on site. The edges of the joint restraining the sealer are usually formed by high-quality concrete or epoxy-mortar nosings or may be fabricated from steel sections. The sealers are now frequently installed by using an adhesive lubricant to assist initial placing and hold the sealer slightly below surface level. This formulation also aids waterproofing, as porosity, surface irregularities and micro-cracking have been found to permit water to pass through the interface between sealer and nosing.

There is no convincing evidence to suggest any marked difference between concrete or epoxy nosings. Either way, it is preferable to allow the surfacing train to cover the joint, suitably protected, and then to saw away that area over the joint to prepare and cast the nosings. In this way a better continuity of the final surface profile is achieved, but with both materials great care is essential in the preparation of the surface. The epoxy formulation should be selected to be insensitive to dampness during construction and should be a flexible mix, not too brittle—for example one containing 15% coal tar. If an epoxy nosing is used, transverse joints should be provided in the nosing using 3 mm ($\frac{1}{8}$ in.) thick expanded polystyrene. These joints accommodate the differential expansion between concrete and the epoxy mortar and should be provided at the rate of about one per traffic lane or every 3·7 m (12 ft). Once the epoxy has set, the polystyrene may be removed by fire or petrol and the joint filled with an approved sealer.

Where steel sections are used, there is no substitute for robustness. Angle sections, for example, when used at surface level, should be at least 13 mm ($\frac{1}{2}$ in.) thick and preferably 16 mm, and should have welded stiffeners at approximately 0·4 m (15 in.) centres if one leg is in the upstand position to contain the surfacing. Any fabricated section should be of 16 mm ($\frac{5}{8}$ in.) minimum thickness and generous anchorage should be provided based on a horizontal design force of 80 kN/m width (2·5 tonf/ft) and a corresponding vertical restraining force of between 400 and 600 kN/m width (12 and 20 tonf/ft). The anchorage bolts should be at least 19 mm ($\frac{3}{4}$ in.) diameter and conservatively rated at about $\frac{2}{3}$ of the proof load.

Many joints have been proved unsatisfactory because of failure of the fixings and, where angles are used to reinforce the opposing edges of the structure, care should be taken to achieve adequate compaction of the concrete under the angles. This requires careful workmanship and, although blow-holes drilled in the upper surface can assist the flow of concrete, it is usually expedient to ignore any contribution from the concrete contained within the angle in assessing the strength of the joint. It is also advisable to provide suitable protective treatment to the holding-down arrangements in this region. Reinforcement used as the cast-in anchorage should be attached to the plates with full-strength welds, at least 19 mm ($\frac{3}{4}$ in.) diameter and designed for forces as above with a minimum of 1600 mm^2/m width of joint (0·75 in^2/ft) at a spacing of not more than 0·38 (15 in.).

An alternative to the cellular preformed sealer consists of a shaped slab of neoprene reinforced with steel plates. Movement is accommodated by shearing strains in the elastomeric material and the steel reinforcing plates enable the joint to span the expansion gap without deflecting significantly under load. The joint is bolted down flush with the wearing surface and is waterproof, although care must be taken with the details at kerbs and edges. It would be interesting to gather details of performance in service of this relatively recent design of expansion joint.

3.5 Joints for large movements

In this range, most joints are of the open type using sliding plates, cantilever or propped cantilever tooth or comb blocks. Water is allowed to pass through the joint and is collected underneath for subsequent disposal. It is extremely important to provide good drainage with no restrictions in which grit and silt can collect. Adequate space and provision should also be allowed for easy maintenance. Because of the passage of surface waters through the joint and splashing in and around the collector system, it is vital that adequate protective treatment should be applied to any parts of the joint exposed to these corrosive elements. The drainage system should be constructed to generous falls as a means of self-cleaning.

With joints of this type, there is always a danger of the free movement being restricted by a build-up of grit and other non-compressible matter between the moving elements. Very often, vegetable growth near the kerbs further consolidates the intrusion and effectively impedes the proper functioning of the joint, imparting high and unexpected loads to the structure, often with disastrous results to both structure and joint.

For movements up to about 130 mm (5 in.), toothed plates can be cut from a plate 40 mm ($1\frac{1}{2}$ in.) thick but for larger movements it is preferable for the comb plates to be cast or fabricated by welding. Castings may be made from either cast steel or spheroidal cast iron: the latter is preferred as it gives less trouble in the casting process.

The design of the teeth should provide adequate longitudinal strength and transverse stiffness to avoid chatter under traffic. The nose of the teeth should also be rounded to accommodate differential deflection, rotation or settlement across the joint. The units should be designed as cantilevers and, whereas some stiffening is useful on the underside, the cantilevers should not be supported at the tips, especially if the structure is supported on rubber bearings or if settlement is expected.

With cantilever joints of this type high stresses are developed under the edge of the supporting unit and, because of this, it is considered desirable to set the seating back from the edge (even though this increases the required capacity of the joint) and to utilize heavy base plates embedded on well compacted granolithic or epoxy mortar rather than use angles cast in the concrete edging (Figure 7b). The holding-down bolts should also be of generous proportions with adequate lever arm and a locking arrangement provided for the nuts. They should be tightened with a torque spanner to about two-thirds the proof loading.

If lever plates are required on comb joints, they should be securely fixed and preferably welded to the teeth at frequent intervals to prevent chatter. Because of chatter, sliding plate joints are not popular for carriageways but they do offer a cheap solution to footway joints if care is taken to keep them free from grit.

The author does not favour the use of a compressed flexible sealer in several units placed back to back.

Throughout this book, the names of proprietary joints have deliberately been avoided but for very large movements the Demag joint would be the author's automatic choice. These joints are robust and of such a size that demands consideration at an early design stage, are installed under the manufacturer's supervision, and have a good service record. Where joints such as these are used, or any other joint which presents a large area of metal on the running surface, the skid risk can be effectively reduced by coating the surface with an epoxy resin mix containing calcined bauxite or similar filler.

The development of modular systems has been reviewed by Watson at the Johannesburg symposium.[23] Developments in this field may prove of lasting benefit in the future.

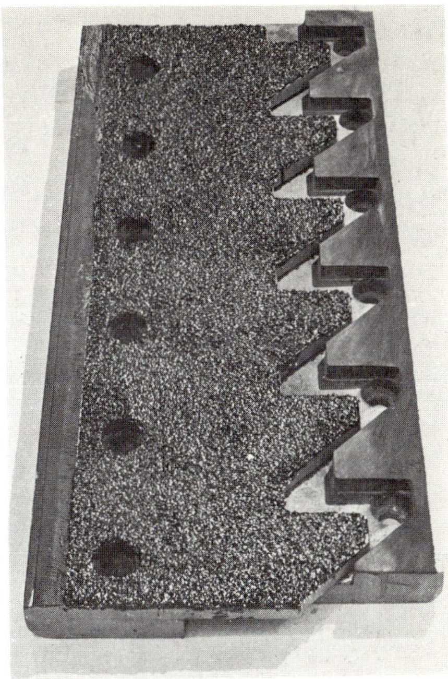

above: Typical comb expansion joint for movement of ± 57 mm (2¼ in.) installed in Bow Bridge flyover, London.

above right: F.T. bridge joint for a skew bridge.

right: Typical F.T. bridge joint for right bridge construction.

below: Typical comb expansion joint for movement of ± 82 mm (3¼ in.) assembled in the factory for the Chiswick–Langley special road near London.

above: View of expansion joint of the comb type, with sliding plate joint at median and edges installed in the Hammersmith Flyover, London. Movement of +134 mm (5·35 in.)−206 mm (8·10 in.)

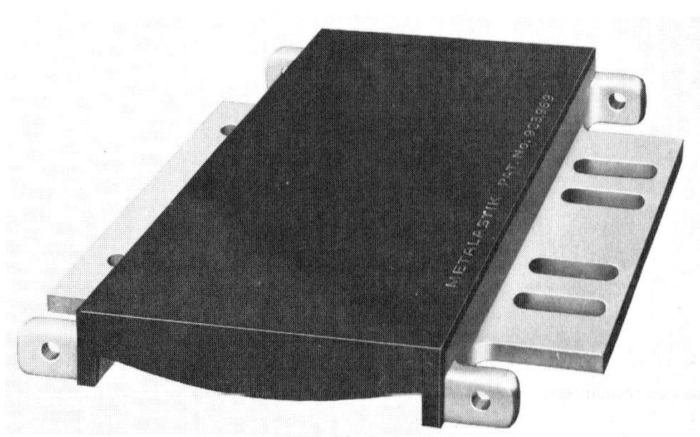

above: View of a typical Metalastik expansion joint for movement up to ±30 mm (1·2 in.)

below: Demag expansion joint being installed in Westway, London.

below: Demag expansion joints in Westway, London, providing for a longitudinal movement of up to 826 mm (32·5 in.).

CHAPTER FOUR

Bridge bearings

4.1 Concrete hinges

Notation for hinges

- a width of throat of hinge (mm)
- D width of member (mm)
- E elastic modulus (N/mm^2)
- f_a stress in edge of throat due to axial load (N/mm^2)
- f_b stress in edge of throat due to rotation (N/mm^2)
- f_y yield stress of reinforcement in tension (N/mm^2)
- H_e effective height of hinge (mm)
- K edge stress ratio
- L length of hinge perpendicular to direction of rotation (mm)
- M moment in hinge (N mm)
- P axial load per mm applied to hinge (N/mm)
- R *Total* resolved component of force acting on hinge $R = L\sqrt{(P^2 + S^2)}$(N)
- S shear force per mm applied to hinge (N/mm)
- U_w cube strength of concrete in throat (N/mm^2)
- ϕ rotation of hinge (radians).

The Freyssinet hinge (see Figure 8a)
The Freyssinet hinge allows rotation of a member to occur by providing a thin flexible throat within the member where large rotations can occur while only small moments are transmitted to the structure. The size of the throat is normally only a fraction of the size of the member but it is able to carry very high axial and shear forces because of the biaxial or triaxial restraining influence of the surrounding concrete. It has been found experimentally that compressive stresses many times the value of the cube stress of the concrete can be applied to a restrained concrete throat without causing crushing, and indeed the primary factor that has to be considered in the design of these hinges is not the compressive stress in the throat but the bursting tension either side of the throat.

Stress distribution in throat
Stresses due to axial load. When the throat of a concrete hinge is subjected to an axial load, stress concentrations occur at the edge of the throat because of the convergence of the stress trajectories in this region. The stress distribution across the hinge under axial load is shown in Figure 9a. Bridle and Sims have shown that if the stress distribution across the throat is parabolic and the stress at the centre of the hinge equals P/D (the average stress in the parent member), then the stress at the edge of the hinge is given by

$$f_a = KP/a$$

where

$$K = 3 - 2a/D$$

Figure 8
Types of concrete hinge.

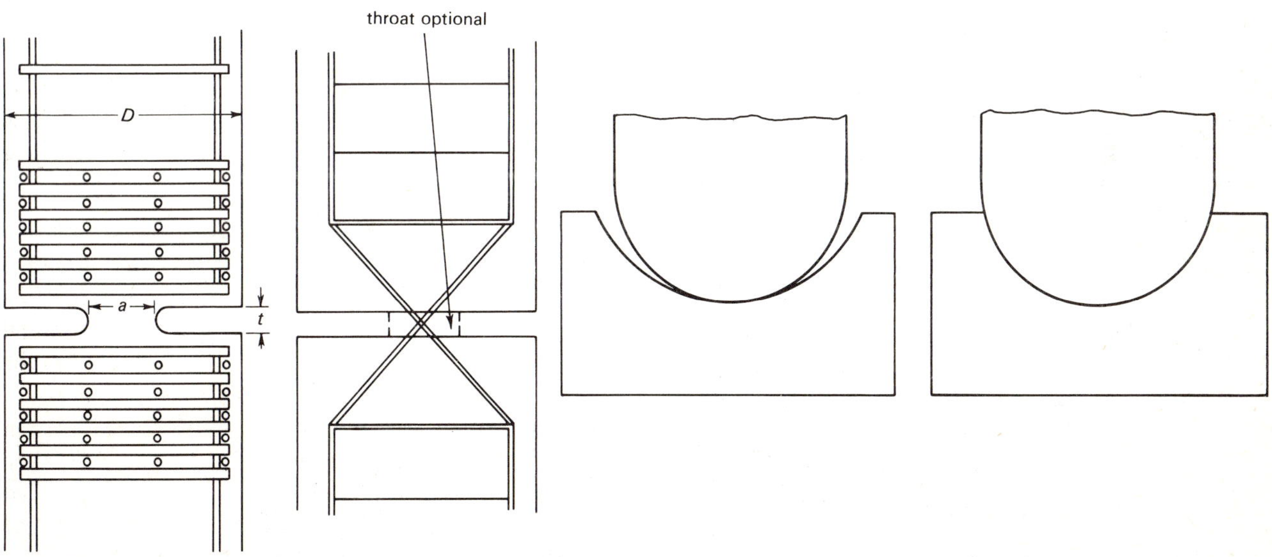

(*a*) *Freyssinet hinge* (*b*) *Mesnager hinge* (*c*) *Saddle bearing* (*d*) *Sliding hinge*

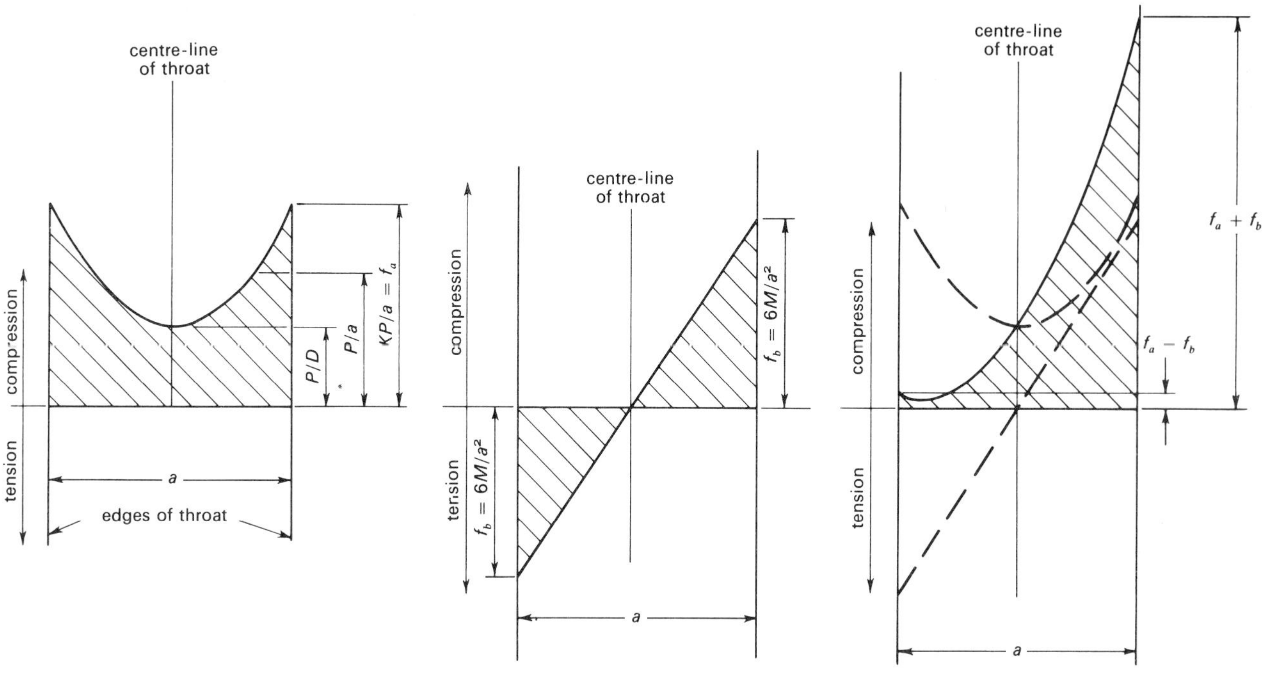

Figure 9
Stress distribution across throat of Freyssinet hinge.

The Ministry of Transport recommends that, when the axial compressive stress at the edge of the hinge is assumed to counteract the tensile cracking stress caused by bending in the throat, a fixed value of $K = 1{\cdot}5$ should be taken.

In no tests carried out so far has a compressive failure of the throat been achieved but it is recommended that the average compressive stress in a throat should be limited to twice the cube strength or 103 N/mm^2 (15,000 lbf/in^2) whichever is less.

Stresses due to rotation. In spite of the high compressive stresses that occur within the throat of a hinge, the triaxial restraint of the surrounding concrete maintains the throat in an elastic state. The bending stresses across the throat are therefore linear and may be superimposed upon the nonlinear axial stresses as shown in Figure 9. If the bending tension at the edge of the throat is f_b then the condition for no tension at the edge of the hinge is given by $f_b < f_a$.

Rotational stiffness of hinge

As the throat of the hinge can be considered to remain elastic during rotation, it can be shown that the moment required to rotate the hinge through an angle ϕ is

$$M = (E\phi)a^3/12H_e$$

where H_e is the effective height of the hinge.

Because of stress flow around the hinge, H_e is always greater than the actual height of the hinge (t). By making assumptions about the shape of the effective M/EI diagram, Bridle and Sims deduced that the effective height of a hinge could be given by

$$H_e = 1{\cdot}33t + 2a(a/d)^{0{\cdot}717}$$

The Ministry of Transport recommend that H_e should be taken as a constant value of 127 mm (5 in.).

If we put the recommended value of $K = 1{\cdot}5$ in the above formulae, we get the condition for no tension at the edge of throat of the hinge as:

$$a \leqslant \sqrt{\frac{3H_eP}{\sum(E\phi)}}$$

where $\sum(E\phi)$ is the sum of the long-term and short-term $(E\phi)$ values.

Bursting tensions

The tensile stresses due to bursting are paramount in the design of Freyssinet hinges. They occur in members adjacent to the hinge over a length equal to the width of the member. They may be calculated by conventional end-block analysis. Extensive tests carried out by the Cement and Concrete Association have shown that the total area of steel required to resist bursting is

R/f_y in the transverse direction,
$R/3f_y$ in the longitudinal direction.

This reinforcement should all be placed within the zone of bursting and as near to the throat as possible. Bond is normally achieved by welding the reinforcement in a series of mats.

Spalling

Although it has been shown that very high compressive stresses can be carried by the throat of the hinge when triaxially restrained, spalling will occur at the ends of the hinge unless the throat is adequately recessed there.

If the notches forming the throat are rectangular in cross-section, spalling will also occur at the corners. Ideally, notches of parabolic profile should be provided but any smooth-profiled end will have a better resistance to spalling than a square end.

Shear across throat
The large compressive stresses present in the throat of Freyssinet hinges give them a high shear resistance. The principal tensions in the throat can be calculated by conventional methods.

The practice of placing shear steel in the throat has been discovered to be of little value. In many cases, in fact, it can have an adverse effect on the hinge.

Where it is required to safeguard the structure against some shear loading of an unexpectedly high intensity (vehicle impact, for instance), dowelling of the member in some position remote from the hinge is preferable to passing shear dowels through the throat.

Prestressing hinges
Uplift may never be permitted in concrete hinges. Where there is danger of uplift, it is possible to prestress across the hinge. The prestressing steel should again be kept away from the throat when possible.

Summary of design requirements
(i) Throat width a
(a) $a \geqslant 50$ mm (2 in.)
(b) $a \geqslant P/103$ (SI units) or $P/15{,}000$ (Imperial units)
(c) $a \geqslant P/2U_w$
(d) $a \leqslant \sqrt{3H_eP/\Sigma(E\phi)}$

If condition (d) is not compatible with conditions (a), (b), (c), the situation may be resolved by prestressing the hinge.

(ii) Throat thickness t

$$\frac{a}{2} > t > \frac{a}{3}$$

(iii) Bursting reinforcement

$$\text{transverse } A_{st} = R/f_y$$
$$\text{longitudinal } A_{sl} = R/3f_y$$

(iv) Shape of notch
The throat of the hinge should be recessed at least 16 mm (3 in.) from all faces of the member. The shape of the notch should be parabolic at the throat.
(v) Rotational stiffness

$$M \text{ applied to structure} = \sum (E\phi)a^3/12\ H_e$$

Mesnager hinge (See Figure 8b)
This hinge is a forbear of the Freyssinet hinge. It carries all the load on reinforcement passing through its throat. In many cases the throat concrete is omitted altogether. The capacity of the Mesnager hinge is considerably less than that of the Freyssinet hinge which has now virtually superseded it.

Saddle bearings (See Figure 8c)
Saddle bearings are used when it is required to accommodate high rotations and shear forces in conjunction with relatively low axial forces. The bearings have the same high resistance to crushing as Freyssinet bearings but they experience the same bursting tendency and should be reinforced with the same amount of steel.

The resistance to rotation of a saddle bearing is small, although the effect of the displacement of the point of contact when the bearing rotates must be considered.

Sliding hinges (See Figure 8d)
These have a similar function to saddle bearings. They are easier to form than saddle bearings but may transmit a significant moment. If the friction coefficient between the sliding faces is μ and the radius is r, then the moment transmitted by the hinge is given by

$$M = Rr\mu$$

(where R is the resultant force acting on the hinge). No rotation of the hinge will occur until this moment has been developed.

Tri-hinge
The tri-hinge shown in Figure 10b is a special application of the Freyssinet hinge. It has a better performance than the conventional Freyssinet hinge because of the multi-directional loading imposed on it.

The tri-hinge is used in conjunction with the rhomboid joint shown in Figure 10a. This joint (which is a concrete version of the Wichert truss support) allows a substantial settlement to occur at a column of an effectively continuous structure, without any change in continuity moment, because the system is statically determinate.

Figure 10

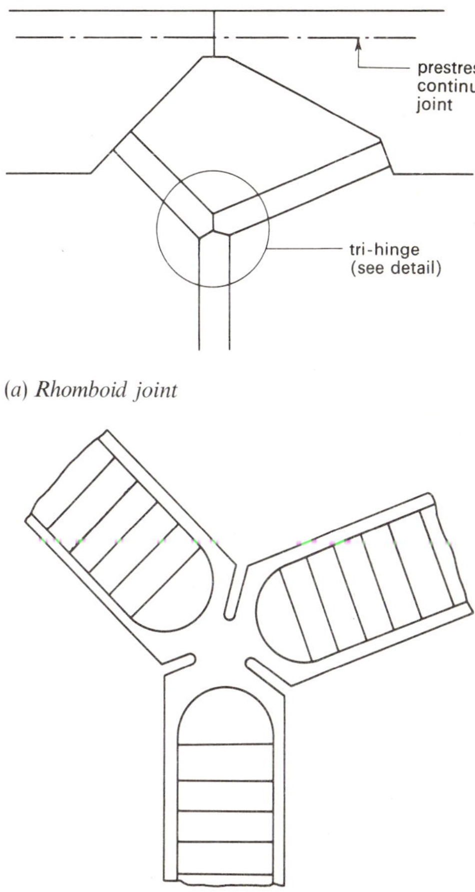

(*a*) *Rhomboid joint*

(*b*) *Tri-hinge*

4.2 Elastomeric pad bearings

The problem of providing bearings that will accommodate comparatively small movements, say 6 to 125 mm ($\frac{1}{4}$ to 5 in.), with or without rotation, has in the past been solved with either some form of roller bearing or sliding plate bearing. Both these types of bearing are not really cheap in first cost and in addition require long-term maintenance which they tend not to receive.

In recent years, bridge designers have turned to elastomeric pad bearings;[28,29] these have advantages of easy installation, no maintenance and low cost. Current experience indicates that, when they are properly designed and manufactured, elastomeric pad bearings have a life expectancy comparable with that of the bridge.

Rubber bearings accommodate horizontal movements by shearing and have enough shearing flexibility to avoid transmitting high loads to the bridge supports. The vertical stiffness is such that significant changes in height under load are avoided.

With experience, it has been found that rubber bearings are extremely versatile and can be used for many unexpected applications by a suitable arrangement.

Design

In Britain, the design of rubber bearings is governed by two documents. For rubber bearings to be used in highway bridges, Ministry of Transport Memorandum No. 802[30] (referred to below as 'MoT Memo') governs. For rubber bearings in rail bridges, the International Union of Railways Code: Volume VII, *Way and Works*[31] (referred to below as 'UIC Code') is used. The UIC Code for rail bridges is of more recent publication and is more detailed in its recommendations. It is included in this volume for reference as Appendix 2. The MoT Memo on the other hand recognizes its own deficiencies and states that future revisions are intended.

Materials

Both the MoT Memo and the UIC Code allow the use of vulcanized natural rubbers and synthetic rubbers. Vulcanized natural rubber must be in accordance with BS 1154: 1952[32] and synthetic rubbers are subject to the engineer's approval.

Some confusion has developed over the relative merits of natural and synthetic rubbers, probably attributable to commercial interests and a lack of specialized knowledge by civil engineers. For instance, natural rubber was excluded from the AASHO specification in favour of neoprene but it has been included since the beginning of 1968. The most satisfactory material used in Britain is vulcanized natural rubber. Synthetic rubbers such as neoprene and chlorobutyl are used, which have the advantage of lower basic material cost but, compared with natural rubber, the ratio of elastic modulus to shear modulus is poor. Also neoprene has a poorer performance at low temperatures. Chlorobutyls have good abrasion resistance and are sometimes bonded to natural rubber fenders in dock work.

Some special properties required of the rubber for use in bridge bearings are that it must have a good resistance to the action of oils, weather, atmospheric ozone and the extreme temperatures to which the bearing is subjected.

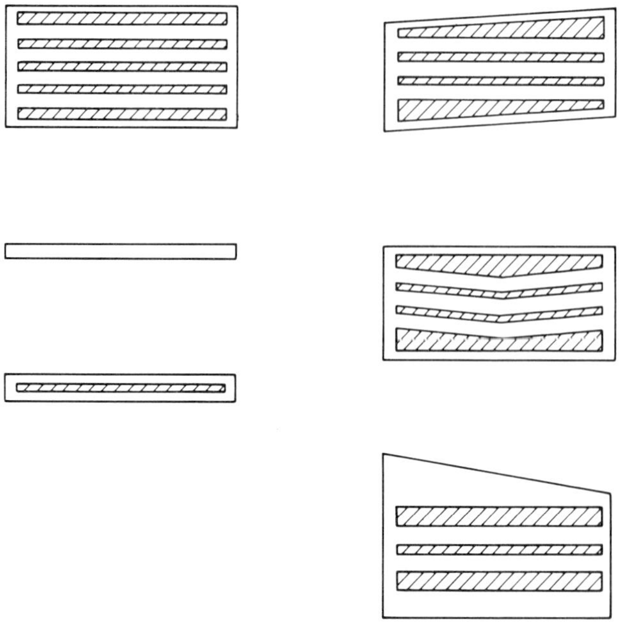

Figure 11
Various pad bearings.

Types and shapes of bearings

There are basically two types of rubber bearing. The first and simplest type is the plain rubber pad or unreinforced bearing consisting solely of rubber. These are generally used to accommodate rotations in locations with little or zero horizontal movement. The second and more versatile type is the laminated or reinforced bearing. These can be designed to accommodate most combinations of load and shear movement, by reinforcing the rubber with one or more steel plates. Sections of these two types together with some other possibilities are shown in Figure 11; the other types shown have been used for particular applications but are now considered largely obsolete and expensive, and in some cases unsatisfactory.

The standard plan shape of bearing is square or rectangular and for most applications this is the most satisfactory shape. However, bearings of circular plan are at present in development and are showing very promising characteristics, the main advantage being that the circular shape has no preferred direction or corners, and in complex structures they can accommodate combinations of shearing forces without the danger of stress concentrations in the rubber. Nevertheless, the rectangular is the simplest to manufacture and is likely to remain the standard type.

Design of bearings

It is common practice for engineers to specify loads, movements, rotations, etc. to a bearing manufacturer who will then prepare designs and quotations. This method gives each manufacturer the freedom to produce what he considers is the cheapest solution. Apparent discrepancies between different manufacturers' designs may be due to the fact that they usually stock a standard range of sizes, which may differ from manufacturer to manufacturer; it is to be hoped that in the future the range of stock sizes will be internationally standardized. The engineer should have sufficient knowledge to be able to check manufacturers' designs and, if necessary, to prepare preliminary designs. It is also of benefit if the engineer understands the design procedure as he is then better able to produce a structure in which the bearings are used most efficiently.

The basic principles for the design of rubber bridge bearings are as follows.

(1) The plan area of the bearing depends on the allowable pressure on the bearing support.

(2) The total thickness of rubber is dependent on the total shear movement and the stability of the bearing.

The MoT Memo gives no guidance on the former condition but ensures stability by limiting the ratio of the thickness of a bearing to its least plan dimension to 1·0 unless the engineer satisfies himself that there is no danger of instability. This ratio is limited to 0·25 for individual layers.

The UIC Code limits the total shear movement to 70% of the total thickness of rubber. Stability is ensured for unreinforced bearings by limiting the ratio of the thickness of the rubber to its least plan dimension to 0·2. For reinforced bearings, the stability condition is dependent also on the dimensions of individual layers of rubber, but it is stated that this condition is generally satisfied if the ratio of total thickness of rubber to its least plan dimension is less than 0·2.

(3) The thickness of individual layers of rubber in reinforced bearings is dependent on the need to limit shear stresses or strains, to prohibit tensile strains, and to limit the vertical deflection.

The MoT Memo gives formulae for calculating the shear strains in the rubber due to vertical load and shear movement only, and it limits both the live load shear strain and the total shear strain to values dependent on the elongation at failure of the rubber. Tensile strains are avoided by limiting the rotation allowable, but the Memo recognizes that this condition is unsatisfactory and states that it gives a 'general indication' only. The vertical deflection is normally limited by considerations of its effect on the structure and is generally not a governing condition in the design of the bearing.

The UIC Code gives formulae for calculating the shear stresses in the rubber due to vertical load, shear movement and rotation. The effects of slowly and quickly applied loads are treated separately, as it has been found that quickly applied loads are more detrimental to the rubber. The total shear stress is then limited to a value dependent on the shear modulus of the rubber. Tensile strain or uplift of the most lightly loaded edge is avoided by conditions which limit the allowable rotation to a third of that allowed by the MoT Memo. These conditions also differentiate between rotations due to slowly and quickly applied loads. Vertical deflections are limited by their effect on the structure and for unreinforced bearings they are limited to 15% of the bearing thickness.

(4) The thickness of the steel reinforcing plates is governed by the need to limit the tensile stresses in them.

The MoT Memo recognizes its inability to give a method for designing the reinforcing plates, but recommends that outer plates should be at least $\frac{1}{8}$ in. (3·2 mm) thick and inner plates at least $\frac{1}{16}$ in. (1·6 mm) thick.

The UIC Code gives a formula for the thickness of the reinforcing plates based on the need to limit the tensile stresses. It is of interest to note that for a bearing with layers of rubber of equal thickness this formula will give thicknesses of the outer plates one-half those of the inner plates.

(5) The protection of the bearing is most satisfactorily achieved by totally encasing the bearing in rubber. The thickness of this rubber casing is at the engineer's discretion, but it will obviously affect the vertical stiffness of the bearing.

The UIC Code states that if the encasing layer (above or below the bearing) has a thickness less than half that of an internal layer and not exceeding 3 mm it may be ignored in design. If the casing layer does not fulfil these conditions, it should be considered as a reinforcing pad of 1·4 times its actual thickness.

(6) The method of attaching bearings must be considered so that slip cannot occur. When no positive attachment is provided, both the MoT Memo and the UIC Code give values for the coefficient of friction to be used for checking that slip does not occur. The UIC Code values for the coefficient of friction are derived from a formula which includes a term dependent on the mean compressive stress.

Layout and positioning of bearings

Bearings must normally be set level to avoid the effects of gravity. An exception to this is when pairs of bearings are inclined to each other to artificially increase their shearing stiffness in one direction.

Although current experience indicates that the life of rubber bearings is comparable to that of the structure, it is advisable to make provision for the replacement of bearings. This may involve providing jacking pockets near each bearing to allow for easy removal and replacement.

If the resistance to longitudinal slipping of the bearing is provided by friction alone, it is advisable to further prevent such slip by means of an anti-creep device.

In structures where shrinkage and creep movements are large, it is sometimes advantageous to provide for the raising of the structure, after some months, to eliminate this deflection.

Specification comments

Materials

The MoT Memo states that vulcanized natural rubber must comply with the requirements for groups Z13 or Z14 of BS1154: 1952 *Vulcanized rubber compounds.*[32]

It is also stated that, if synthetic rubbers are used which have properties differing from those laid down in BS1154, variations may need to be made to the basis of design, permissible stresses and workmanship given in the MoT Memo. The UIC Code merely states that the rubber must comply with any regulations and standards in force. It is therefore advisable to specify that the properties of the rubber comply with BS1154.

The properties required of the rubber can be split into two classes. The first may be termed 'physical properties' and these include resistance to the action of oils, age, weather, atmospheric ozone, ultra-violet radiation, and extreme temperatures. The second class may be termed 'mechanical properties' and these include requirements governing the hardness, tensile strength, elongation at break, and compression set.

The mild steel used for reinforcing plates should comply with the requirements of BS15 or BS970.

Testing

The nature of the testing required falls into two classes: firstly, tests on the materials used to verify their specified properties; secondly, tests on complete bearings to verify the adequacy of the design and manufacture.

The tests on materials may be further subdivided into those that verify the material's 'physical properties' and those that verify its 'mechanical properties'. Tests that verify a material's 'physical properties' are inherently long-term and little can be done at the time of manufacture. The tests specified in BS1154 are limited to accelerated ageing and low-temperature flexibility tests. The MoT Memo also specifies a weathering test which lasts eight weeks. In addition, if natural rubber is used, limits on the chemical composition ensure the required properties. However, if synthetic rubber is used, the specifying engineer is in the slightly invidious position of having to rely on a reputable manufacturer to provide a suitable material.

Tests to verify a material's 'mechanical properties' are those laid down in BS1154 and described in the relevant part of BS903.[33] In addition, the MoT Memo specifies a test of the bond between the rubber and the steel reinforcing plates. Tests on the steel for the reinforcing plates are those carried out normally on steels to BS15 or BS970.

Tests on complete bearings are limited to verification of their shear and compression stiffnesses, although the UIC Code also specifies an ultimate compression load test.

The amount of testing to be done is at present the subject of some confusion. The MoT Memo leaves the decision to the engineer and so could depend on the engineer's confidence in the manufacturer. The UIC Code requires that a full series of tests should be carried out during manufacture on three samples taken from each batch of 150 bearings. For orders of small quantities of bearings, the engineer must presumably rely on past test data supplied by the manufacturer. For the acceptance of a quantity of bearings less than 20 in number, the UIC Code requires that all bearings must be tested in compression and half must be tested in shear. For larger quantities, additional bearings must be made from which samples can be cut and subjected to a full series of tests.

Workmanship, inspection and maintenance

It is important to avoid corrosion of metal reinforcing plates and this is most satisfactorily achieved by totally encasing the bearing in rubber. This also serves as an additional protection to the rubber.

Metal reinforcing plates should be free from sharp edges.

The use of welding for bonding metal-to-metal surfaces should not be permitted where there is a danger of burning the rubber or affecting the bonding of the bearing. Adhesives used for bonding metal-to-metal surfaces are generally high-viscosity resins which set cold and are solvent-free.

Tolerances for the thicknesses of single rubber layers and the total thickness of the bearing are specified in the UIC Code. However, it is normal practice for engineers to accept manufacturers' tolerances subject to visual checks made during the testing of complete bearings.

Inspection of bearings in service should be carried out from time to time to confirm that their performance is satisfactory. Any steel parts exposed must be protected from corrosion by periodic painting or other means. It is also important to ensure that the bearings are kept reasonably clean, particular care being taken to avoid contact with grease, oil, petrol and similar substances which might attack natural or synthetic rubber.

4.3 Pot bearings

The problem of providing a bridge bearing to take large vertical loads combined with a high degree of rotation cannot be overcome by using laminated rubber bearings; they would require large bearing surfaces to keep all sides of the bearing in compression and this would result in an uneven pressure distribution on the concrete, which in most cases would be too great. In the past, this problem was overcome by using heavy steel rocker bearings.

The pot type of bearing was developed in 1959, consisting of a circular non-reinforced rubber or neoprene pad, of relatively thin section, which is totally enclosed by a steel pot. The rubber is thus prevented from bulging and has been shown by extensive testing to behave similarly to a fluid under high pressures. The result produces a bearing suitable for rotations of up to $\frac{1}{50}$, at the same time giving the effect of a point-contact rocker bearing since the centre of pressure does not vary by more than 4%. This type of bearing can be combined with a sliding or roller bearing to give horizontal movement (see Figures 12 and 13).

Experience has shown that the pot bearing is a compact, economical and efficient bearing. The only two points of possible doubt are firstly, the ability of the steel pot to withstand high transverse loadings combined with the pressures from the rubber, and secondly the fact that satisfactory rotational operation of the bearings is not achieved until at least 25 % of the working load is applied, so that care must be taken in specifying erection procedures.

4.4 Steel bearings

Recent developments in bridge design, especially the use of large prestressed concrete members, have resulted in bearings being required to take even higher loads while at the same time accommodating large movements and some degree of rotation. Conventional steel bearings, if adapted to these conditions, would require a nest of rollers to accommodate the vertical load and horizontal movement, surmounted by a rocker bearing (or in some cases a knuckle with a large-diameter pin) to accommodate the rotations. If one considers a 100 mm (4 in.) diameter pin of mild steel to BS15, a permissible load of 2 tons/in. can be taken on a design in accordance with BS153. For a load of 400 tons, this would require 10 rollers each 20 in. long.

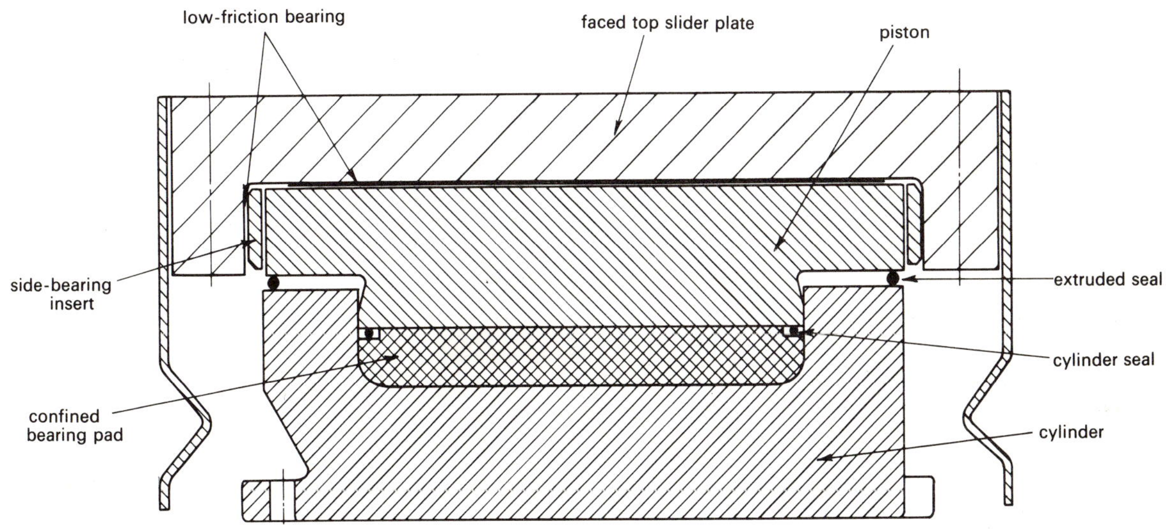

Figure 12
Metalastik bridge bearing: constrained rubber type as used for Westway, Section 6.

Figure 13
Bearings for Mancunian Way.

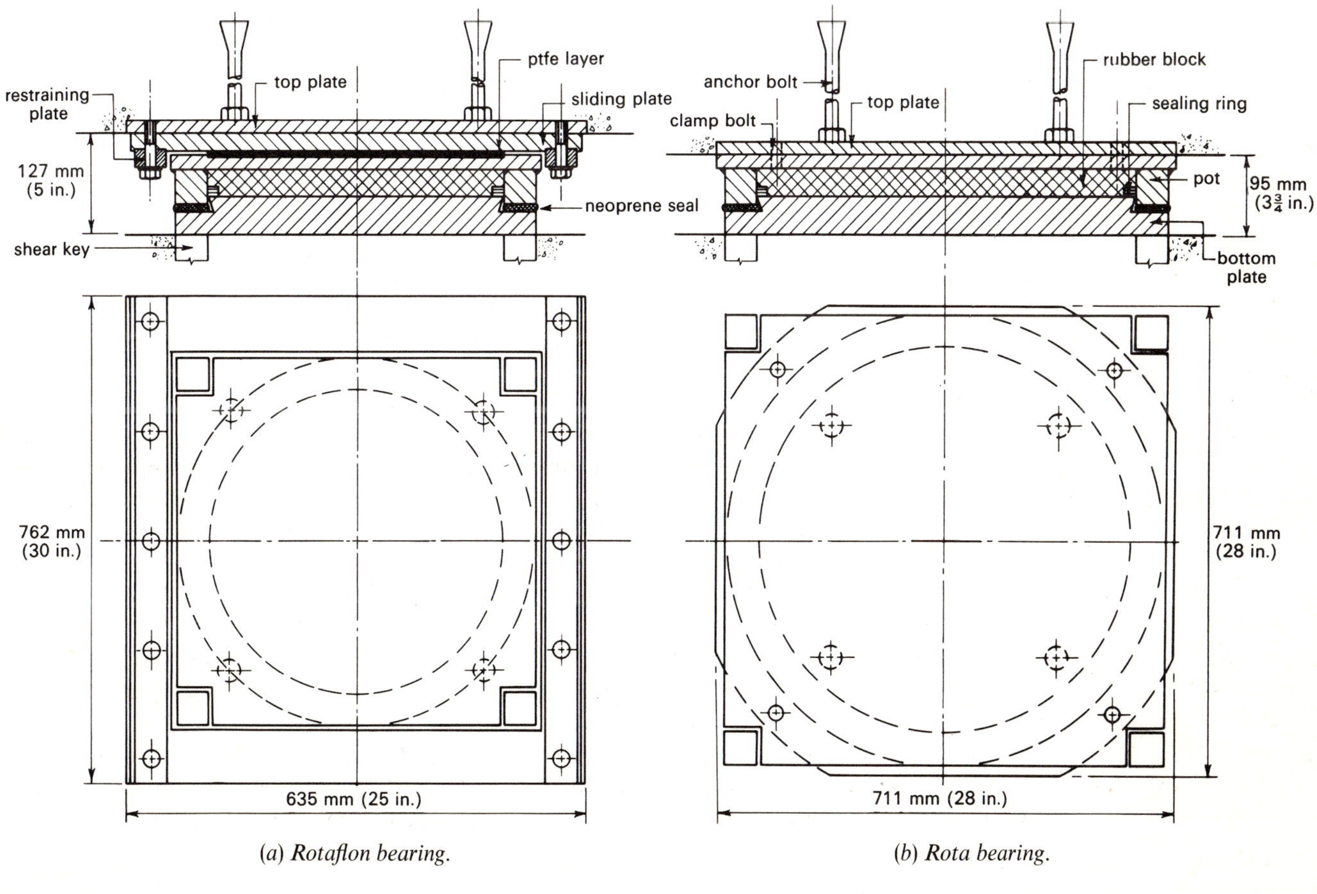

(*a*) *Rotaflon bearing.* (*b*) *Rota bearing.*

The solution to this problem depends on increasing the allowable bearing pressure or Hertz pressure between a circular roller and a flat plate, which is largely proportional to the surface hardness of the steel. At the same time, it should be emphasized that this should not just be a surface-deep hardness which would tend to crack under service. The results achieved have meant that a single roller can be used, thus eliminating the necessity for a rocker or knuckle. For these improved steels, an allowable load of 10,000 kN/m (25 tons/in.) can be achieved for a single 100 mm (4 in.) diameter roller, which will take 4000 kN (400 tons) over a 400 mm (16 in.) effective length.

Three basic types of roller are currently produced, and a brief comment on their different performances demonstrates what has been achieved. The first hardened steel bearings to be manufactured in Germany were the Kreutz Armour steel bearings of high-grade stainless alloy steel, which are hardened to give a Brinell hardness of 600 for up to 13 mm ($\frac{1}{2}$ in.) penetration, and under test conditions a coefficient of rolling friction less than 1%. The alternative German system developed by Esslingen uses a 'Corroweld' surface consisting of multi-layer welds build up on the roller and bearing plates and then machined to give a smooth bearing surface. The Westwood Hi-load system used a high-grade chromium alloy steel, heat-treated to give a Brinell hardness of 500 for up to 32 mm ($1\frac{1}{4}$ in.) penetration. The design working load is 96·5 MN per m diameter per m length (6·25 tonf per in. diameter per in. length) and under test conditions the coefficient of rolling friction is less than 0·2%. In practice, tolerances in setting the bearing plates and allowance for long-term surface deterioration indicate a design coefficient of rolling friction of 3% might be prudent.

For heavy loadings, economic fixed-rocker bearings are made using these same three systems and the most satisfactory results are produced by the Hi-load material. For both types of hardened steel bearings, the design considerations are the same, two typical examples being shown in Figures 14 and 15.

The question of adequate protection for the hardened steel surfaces, and hence the long-term performance of the bearing, should be carefully considered. The most satisfactory arrangement is to enclose the bearing in an easily dismantled grease-box, as in the Westwood system.

In one project in Australia, the authorities have preferred to specify a hard chrome-plating protection, and hence keep the roller surfaces visible. A note of caution should be stressed regarding chrome plating. If the plating is too thick, it will cause cracking which results in a surface that is inferior to an unprotected one. If the plating is too thin, there is a tendency for small spots of rust to form because of the effects of hydrogen pitting. A satisfactory thickness for hard chrome plating is 0·06 mm (0·0025 in.), but it is thought that the best form of protection under varying conditions is afforded by a grease-box.

Since the roller moves very slowly, it has been found best to use a grease which has the same consistence as putty, so that there is no danger of leakage causing unsightly stains, but at the same time the grease must not be so thick that it parts from the roller surface. The Westwood system uses Shell Barbatia 4, a graphite-based grease.

A recent development in Westwood Hi-load bearings is to omit the grease boxes and substitute stainless steel for the Hi-load parts. From various tests that Westwood carried out, a satisfactory design has been evolved and the design working load is 77·2 MN per m diameter per m length of roller (5 tonf per in. diameter per in. length).

4.5 Sliding bearings

Introduction

Sliding bearings of a metal-upon-metal type have been in use for more than a century, generally bronze sliding against steel or cast iron with some sort of lubrication. These were very simple and cheap, but even when freshly and evenly lubricated the coefficient of friction is of the order of 20%. It was also found that, even with regular maintenance, the coefficient of friction increased to unacceptable limits. However, at a time when it appeared that sliding bearings were to become a historical curiosity, they were revitalized by the introduction of polytetrafluoroethylene, ptfe.

Ptfe sliding surfaces

Ptfe is a fluorocarbon polymer, one of a group of plastics remarkable for their extreme chemical resistance, excellent dielectric properties and wide range of working temperature. They were first produced, other than on a laboratory scale, during the Second World War for use in the handling of highly corrosive products connected with the manufacture of uranium 235. After the war, ptfe was put on the open market and the advantages of its low coefficient of friction were realized.

Ptfe was found to have the lowest coefficients of static and dynamic friction of any solid; they are numerically equal and this leads to a nominal absence of stick-slip movement. However, pure ptfe has a low compressive strength, high thermal expansion and very low thermal conductivity, and is therefore not immediately suitable for heavy bridge bearings. However, these detrimental properties can be improved by the use of fillers, without affecting the low-friction properties. The most common fillers are glass-fibre and bronze. With this development, bridge bearings incorporating ptfe-based sliding surfaces became practical and were first used at the beginning of the 1960s.

There is now available a variety of ptfe-based bearings, the low-friction surfaces being used either to provide rotation by sliding over cylindrical or spherical surfaces, or to provide horizontal sliding movement, or a combination of both. To design engineers, the factor which is of greatest interest is the coefficient of friction, this being a critical aspect in the design of structures incorporating ptfe sliding bearings.

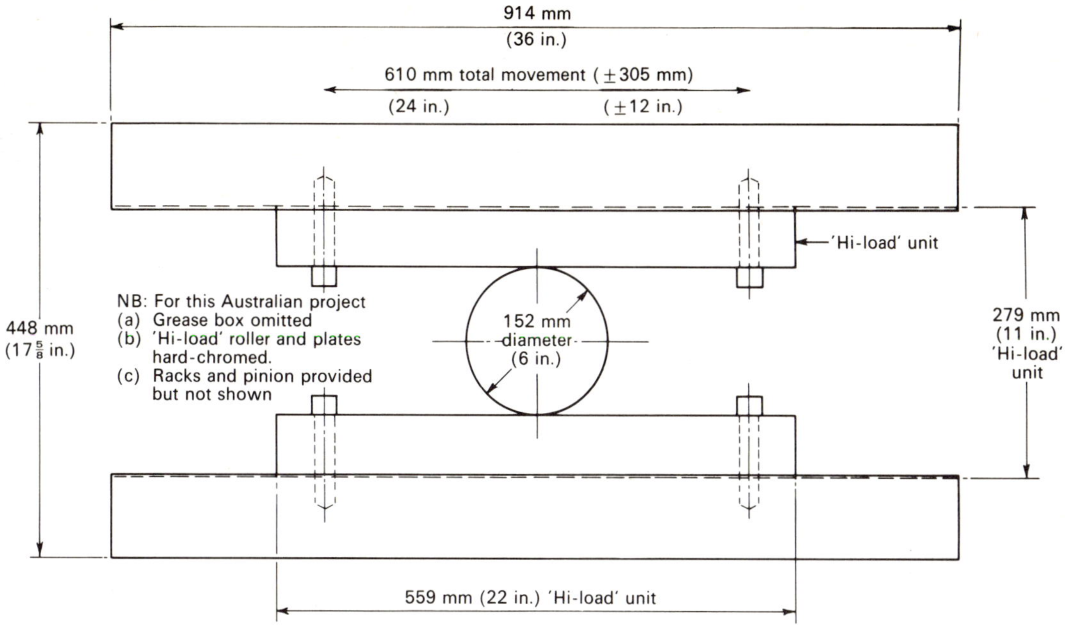

Figure 14
900-tonne Westwood 'Hi-load' roller bearing for West Gate Bridge.

Figure 15
1400-tonne Westwood 'Hi-load' rocker bearing for West Gate Bridge.

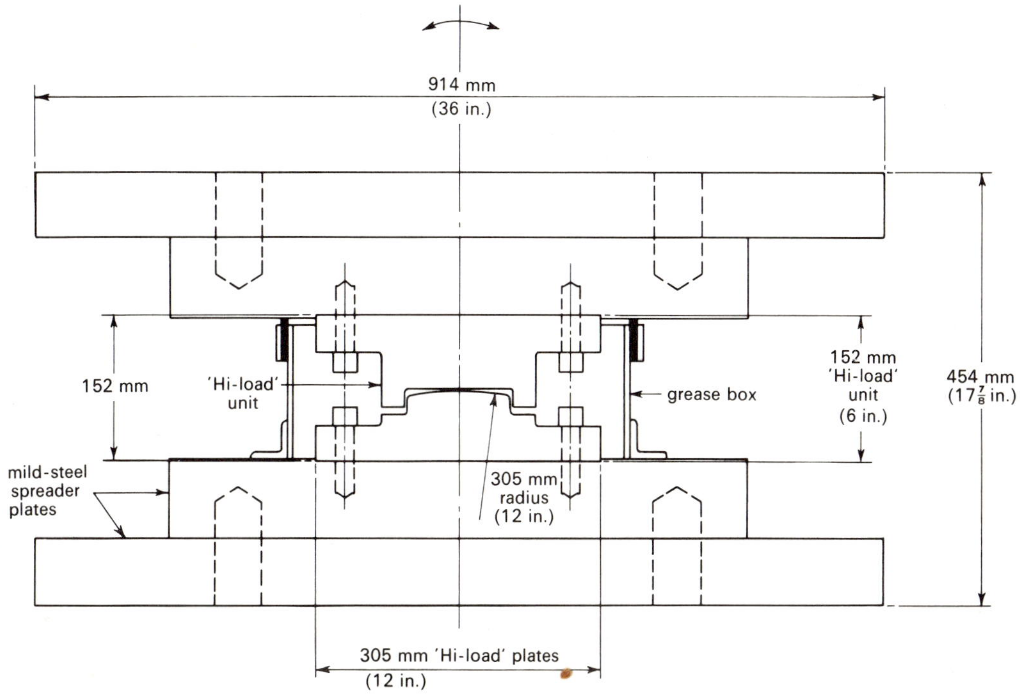

There are three methods of arriving at a value for the coefficient of friction:

test specimens of the material in a laboratory;

test complete bearings in a laboratory;

test complete bearings under service conditions when incorporated into a structure.

This is a logically developing sequence and has been followed in practice. The first method, testing specimens of the material, has been carried out at Stuttgart University and the Road Research Laboratory. These tests show that the friction coefficient can be below 1% given appropriate conditions.(34) The second method, testing complete bearings, has been carried out by manufacturers, and the latest information on this(35) arrives at a coefficient of friction of 3% which will reduce with age. The third method, site tests on bearings in place in a structure, has been carried out on a number of sites. The most direct method of doing this is to jack an entire deck and, from a knowledge of the self-weight and jacking force, calculate the coefficient of friction. The values arrived at by this method vary between 2 and 4%.(36) An alternative method is to measure strains set up in parts of the structure by the friction force and relate them to its value. This method was used for Coldra Viaduct, South Wales, and values for the coefficient of friction between 3·9 and 4·5% were found.

Fabreeka bearings

The Fabreeka bearing is a relatively new type employing reinforced rubber in the shape of a rectangular block purely as a rotation element and with ptfe bonded to the rubber for sliding action. Reinforcement to the rubber consists of multiple layers of compressed cotton duck. The 0·8 mm ($\frac{1}{32}$ in.) layer of filled teflon is bonded to the Fabreeka and a stainless-steel plate is attached to the underside of the beam by epoxy-resin adhesive or rag-bolts.

The author has no experience of the performance of this bearing. It is intended as an economical bearing for the lighter and smaller applications.

Friction values of ptfe for design of structures

The frictional coefficient of ptfe sliding bearings appears to be relatively high when the bearing is new and may even increase when it is bedding down, but reduces with time. It is also reduced at the very slow rates of movement which apply in bridges. Accelerated tests have not shown that wear in bridge bearings is likely to cause an increase in the frictional resistance, at least not in the operational life of the bearing.

It is necessary to design a small group of bearings for maximum friction. In particular, any structural element should be checked against an increase in friction so that any overload can be safely taken within a permissible overstress. Since the forces are proportional to the friction coefficient, it is natural to aim at reducing it as far as possible but it is important to establish accurate values. When considering friction alone on a small number of bearings where ptfe is rubbing against stainless steel, a coefficient of 5% has been adopted by the author; 6% might be appropriate for certain types of filled ptfe. With a large number of spans in a continuous structure, the statistical probability of all the bearings carrying a maximum friction load is reduced and it would be reasonable to design for an overall value somewhat lower. This is relevant when one considers friction on all the bearings combined with, say, wind and centrifugal forces. A combination of forces might take account of the maximum centrifugal effect developed from vehicles travelling at full speed over a given length of structure, and vibrations from fast-moving traffic which would tend to promote free slip action; it would also allow for the fact that traffic is unlikely to be travelling fast when wind speeds are high or the extremes of temperature are being experienced, and that maximum wind pressure is unlikely to affect the whole length of a long structure at a given instant of time.

Under these effects, an anchorage may be designed to withstand the sum of the 5% bearing friction. If an anchorage is located approximately in the centre of a long structure, the difference in friction occurring between the two opposing lengths can be taken into account. One might take 7% from one half, counteracted by 3% from the other half. When looking at the forces in friction in combination with the maximum effects of wind and centrifugal force, 3% has been taken and it may even be possible to reduce it to 1%. It is important to relate the friction value being taken to precise loadings occurring in combination with other forces as it cannot be stressed too often that the value of the coefficient of friction changes with contact stress and other factors.

When large horizontal forces are being accommodated, one has to consider whether provision of ptfe rubbing surfaces is necessary to reduce the forces generated. It can be seen that, with a vertical sliding friction value of 5% and a steel-to-steel rubbing contact for lateral loading of say 25% to 75%, one can arrive at an incompatible situation of loading on the substructure.

Shift of centre of pressure in ptfe bearings

There will be a slight shift in the centre of pressure because of the frictional resistance of curved sliding surfaces which allow rotations. This shift will be equal to the radius of curvature of the sliding surfaces multiplied by the coefficient of friction, and is generally small. However, the radius of curvature of the sliding surfaces can be large—for instance, values in excess of 1 m—and in this case the shift of the centre of pressure can be more than 50 mm if 5% friction is assumed.

Setting tolerances for ptfe bearings

One aspect that requires attention is the tolerance permitted in the setting of sliding bearings.

Suppose a sliding bearing is to be set truly horizontal but is tilted in the direction of sliding by 10 mm in 1 m ($\frac{1}{8}$ in. in 1 ft). This inaccuracy is responsible for generating an additional apparent friction force of about 1% when the structure slides uphill—a high proportion of the true sliding friction for which proper allowance has therefore to be made.

A similar effect becomes apparent when one considers a change in rotation of the bridge deck relative to the supporting structure. Rotations may arise from any loading producing flexure such as prestressing forces or flexibility of the substructure.

Preferred positioning of curved sliding surfaces

When ptfe is employed on curved surfaces, attention should be paid to the optimum geometrical arrangement to reduce movement and therefore wear on the sliding surfaces. Figure 16 illustrates such a consideration.

Balancing torsional stresses with eccentric bearing reactions

In his paper to the Johannesburg Symposium, Bobisch(37) mentioned an interesting application for some tightly curved bridges on the Durban Freeway system. In normal cases, the ptfe sliding surface is above the rotation element and, during sliding, an eccentric reaction is produced at the soffit of the superstructure. Structures which are curved to radii less than 100 m (300 ft) are sensitive to torsion. The Williams Road ramp at Albert Park, Durban, of 35 m (115 ft) radius, was designed with the bearings placed in the inverted position, since the effect of the eccentricities of the bearings which all act in the same direction is cumulative and could have led to very high torsional stresses in the deck. The solution

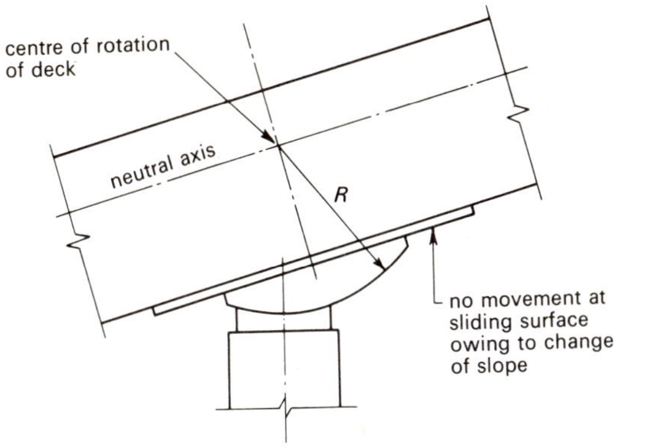

(*a*) *Where centre of curved surface is coincident with centre of rotation of deck*

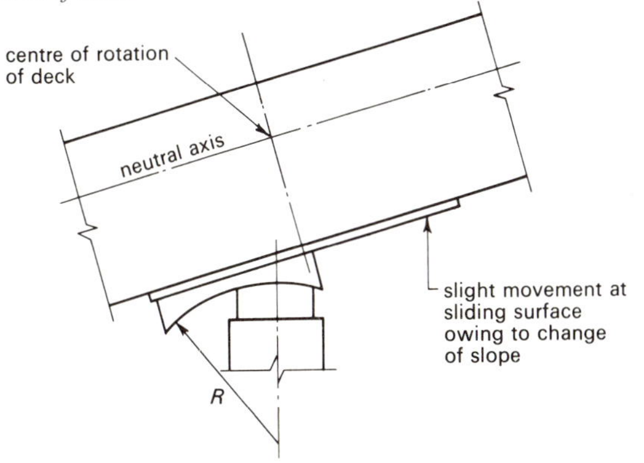

(*b*) *Where centre of curved surface is not coincident with centre of rotation of deck*

Figure 16
Effect of rotation.

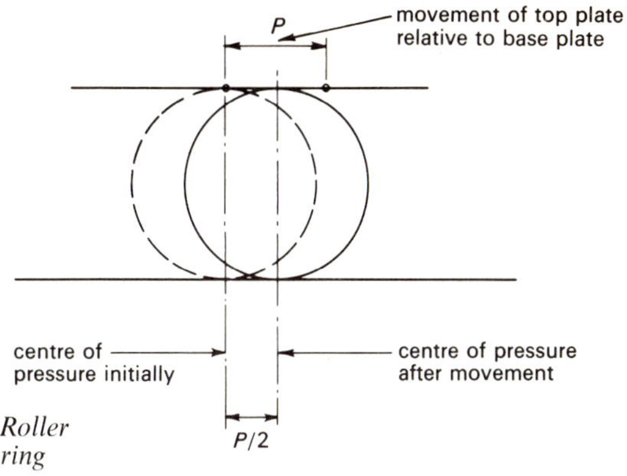

(*a*) *Roller bearing*

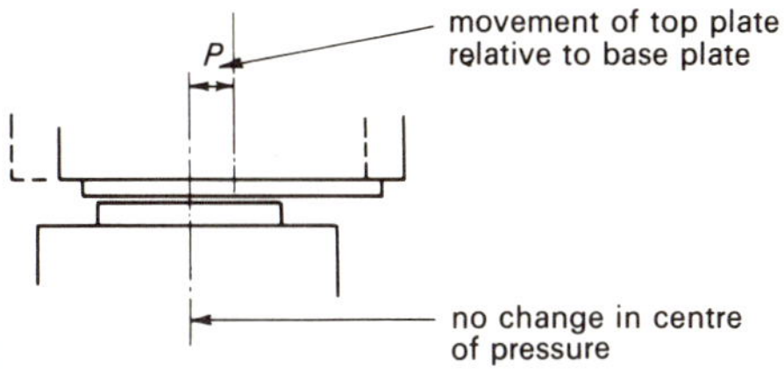

(*b*) *Sliding bearing*

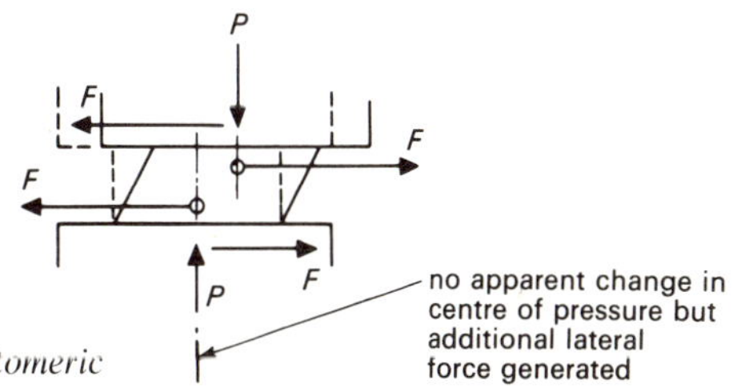

(*c*) *Elastomeric bearing*

Figure 17

adopted transferred the resulting eccentricity to the individual supports where they could be easily accommodated.

4.6 Comparative selection of types of bearing

A big advantage that sliding bearings have over bearings of the roller type is that they can be arranged to obviate eccentricity of loading (Figure 17). For a given movement it can be seen that a roller bearing inevitably provides an eccentricity of load from its mean position of half the movement. On the other hand, if the sliding plate is built into the moving structure and the rotational element is built into the substructure, the loading remains axial in spite of movement.

Modern designs of bridge require the simultaneous treatment of both transverse and longitudinal movements. For bridges of varying width, skew and curvature, care must be taken when laminated rubber bearings are used to ensure that the movements to be expected are properly evaluated to prevent overstressing of the bearing. In certain circumstances when large movements are involved, it may be preferable to use sliding bearings which can allow unrestrained movement in certain directions and inhibit movement in other directions by guide plates.

The designer has to distinguish clearly between the characteristics of bearings of various types and evaluate correctly their parameters as part of the structure. Elastomeric bearings have a force-displacement relationship which generally enables them to be included in the elastic analysis of the structure. For roller bearings, we have to consider first the force required to set the roller in motion; once it is rolling, a nominally constant friction force is involved.

For sliding bearings, a somewhat similar set of circumstances applies. Movement of the structure, if small and below the friction force of the bearing, is absorbed by the elasticity of the structure. Thereafter, a greater movement builds up a force equal to the sliding friction and the force then remains substantially constant although dependent on such factors as speed of sliding and contact stress.

For modern sophisticated structures, it may not be valid to change the type of bearing or even the manufacture without considering the effect on the rest of the structure.

4.7 Specifications for bearings

The method used for selecting and specifying bearings for Western Avenue Extension in London is of much interest. Separate sub-contract documents were prepared which included clauses relating to materials, workmanship and any particular requirements. The most important part of each document was the schedule giving all loadings and movements with which the manufacturer had to comply in his design and quotation.

Particular points which were borne in mind in drafting the specification included the following.

(1) Bolts to be vibration-resistant.
(2) Adequate specification of tolerances to be given both for machining and the finished bearings.
(3) The nature of any protection should be included.
(4) Whether keeper plates are to be removable.
(5) What overall height and size restrictions apply.
(6) How the bearing is to be fixed to the column, plinth and superstructure.
(7) Whether the insertion of shims to allow for settlement or tilt will be required.
(8) Clear and realistic specification of proof and working load testing, particularly when bearings of high load capacity are required, as testing facilities may not have sufficient capability.
(9) Whether the clamping of parts of bearings together for transit and placing is required, and whether allowance is to be made for offsetting bearings in the factory ready for use on the site.
(10) Whether folding wedges or templates for bearings are required to be supplied before bearings are despatched from the manufacturer.

Tenderers should be asked to provide calculations and general arrangement drawings of their proposals to meet the specified requirements at the time of quoting. Where progress in design is advancing so rapidly, the engineer has to use technical judgement in assessing quotations and not necessarily accept the lowest tender. In assessing designs, attention might be paid to such matters as:

the thickness of the main upper and lower plates to give adequate resistance to twist;
the general layout of welding details if incorporated to avoid distortion problems during manufacture;
the type of special materials used, such as ptfe, and the design bearing pressures (these are generally increasing in line with the trend of competition and progress in manufacture);
general maintenance features which the manufacturer may otherwise overlook, such as ease of removing clamping plates or bolts when the bearing is incorporated in the moving structure, together with the general ease of removal of the complete bearing if required in the future.

Some particular features of the bearings used in the Western Avenue Extension are mentioned in Chapter 5. The cost of the bearings involved in the sub-contracts is nearly 2% of the cost of the main contract.

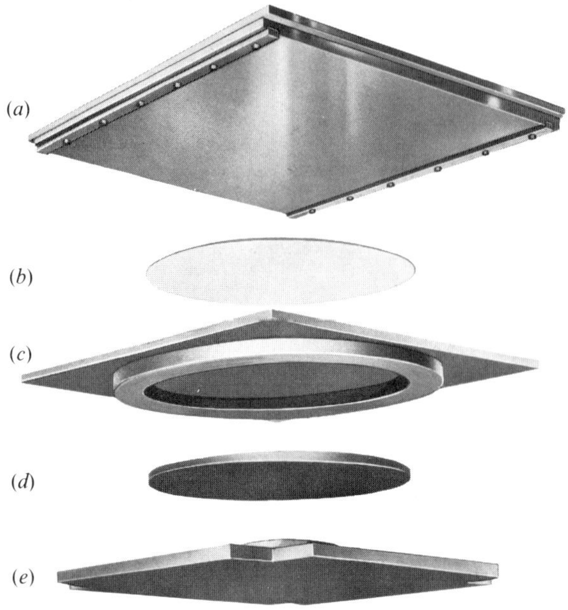

Exploded view showing components of Rotaflon bearing.
(a) Stainless-steel faced sliding plate (b) Teflon (ptfe) disc
(c) Cylindrical ring plate (d) Rubber disc
(e) Plate with integral piston.

Concrete tri-hinges incorporated at the top of the piers in a footbridge on the M1 Motorway in the West Riding of Yorkshire.

right: A double concrete hinge bearing under test at the Research Station of the Cement and Concrete Association. The hinge is supporting a load of 58 tons under a displacement of 85 mm (3·35 in.).

below: P.S.C. laminated rubber bearings on top of the columns in the Huntley's Point Overpass, Sydney, Australia.

right: Testing rubber bearings in compression and shear in the Andre Rubber Company's factory test rig.

below: Some partly finished Rotaflon bearings in the workshop. Illustration shows 850-tonne special Rotaflon bearings with removable guides for Westway, Section 1.

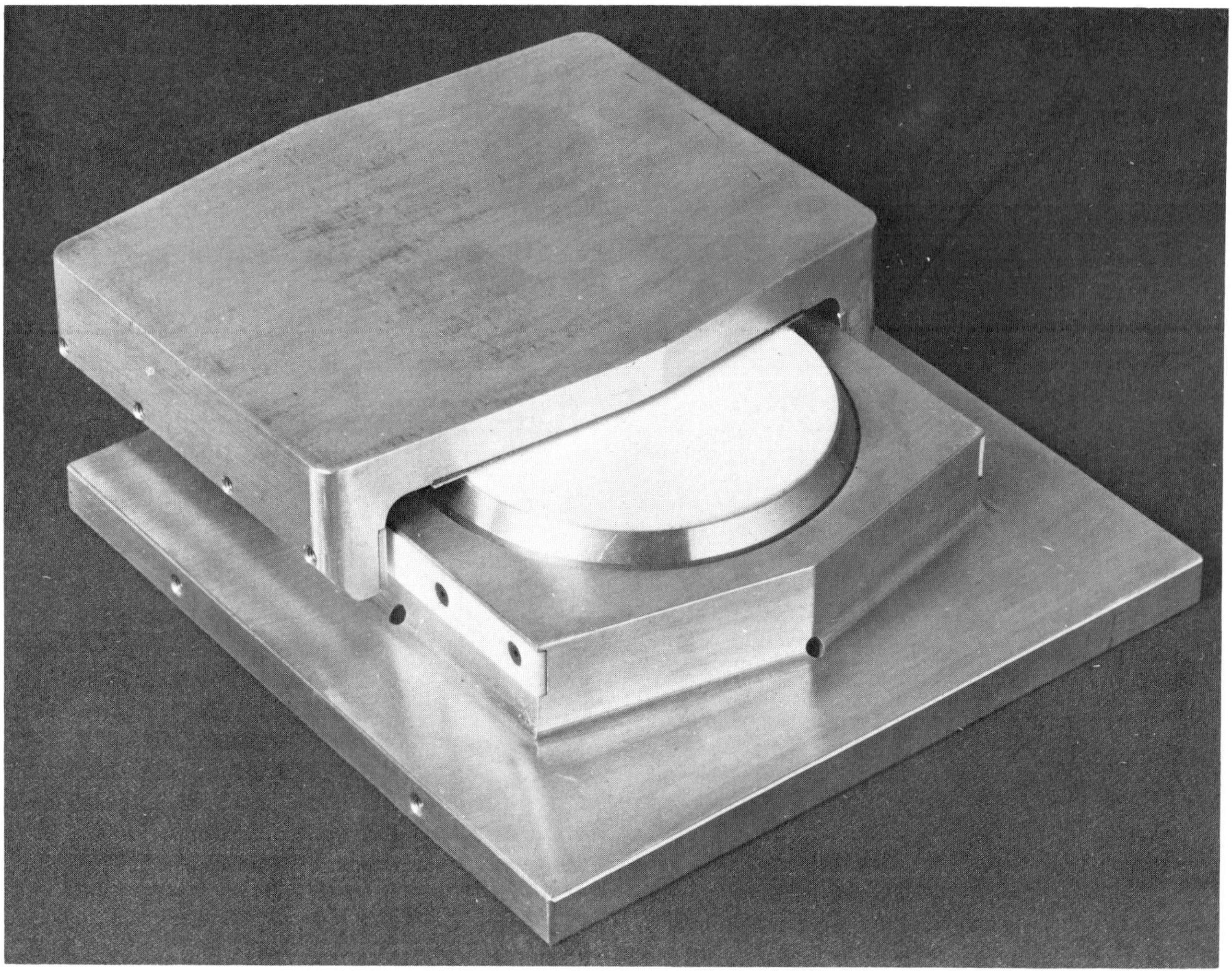

above: View of Solarbridge floating piston bearing.

left: View showing floating piston bearings being installed for the Thames Bridge at Chertsey on the M3 Motorway.

right: Westwood Hi-load 3000-tonne spherical bearings for the Calder Bridge on the M1 Motorway.

below right: Adur Bridge, Shoreham Bypass. Double mild-steel rocker bearings combined with double Hi-load roller bearings (two rollers per bearing) on skew. Total capacity 1200 tonnes.

below: Westwood Hi-load roller bearing of 370 tonnes capacity with provision for ±330 mm (13 in.) of movement for a railway underbridge on the Sheffield–Leeds Motorway.

right: Westwood Hi-load roller bearing of 600 tonnes capacity for Mersey Tunnel flyover.

below: West Gate Bridge, Melbourne. Hi-load rocker bearing of 2800 tonnes capacity. Rocker components of high load steel. Inner and outer plates of mild steel.

above: Hi-load sliding rocker bearings for Dungeness Power Station.

left: 3200-tonne spherical Hi-load batches for Almondsbury Interchange. Bottom rocker assembly.

above: View of precast beams being placed on Tetron cylindrical bearings.

left: View of Tetron cylindrical bearings in position for a right bridge on the M1 Motorway.

below: View of Tetron cylindrical bearings in position for a skew bridge.

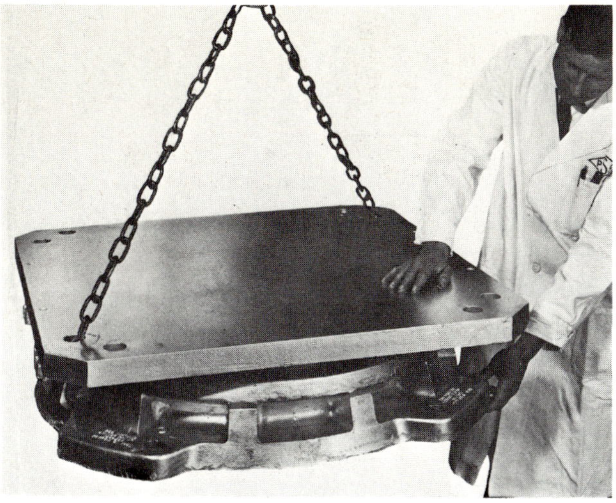

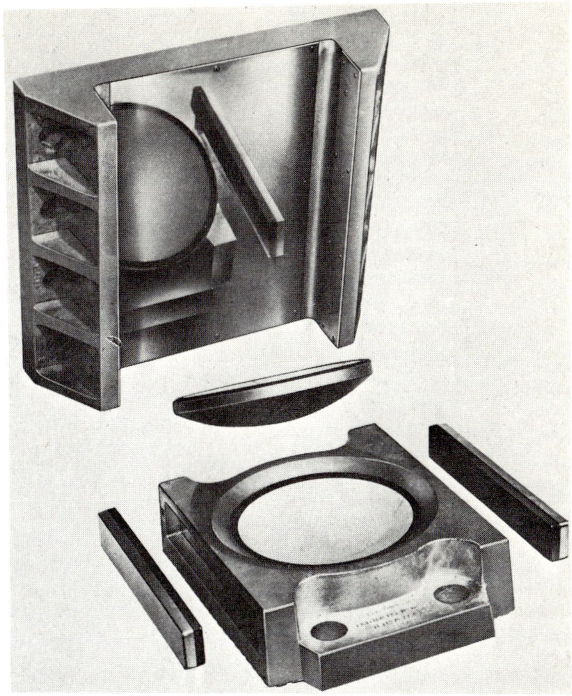

above: View of Tetron cylindrical bearing for bridge structure on the Johannesburg Ring Road.

right: Tetron SFJ bearing providing movement in one direction only and providing restraint in the direction at right-angles of 50 % of the maximum load rating independent of the direct load on the bearing.

above: View showing Glacier bearing in position in Westway Section 5, London.

right: Assembly of Glacier anticlastic bearing for Westway, Section 5. Working load up to 2770 tonnes. The anticlastic sliding surfaces are faced with DU material.

below: View of Glacier anticlastic sliding bearing as assembled.

above: Westway, Section 1. View of Andre Rotaflon bearing at top of pier with flexible cover on bearing.

left: Westway, Section 1. View of structure showing Andre Rotaflon pot bearings at top of piers.

below: Metalastic sliding bearings in position at foot of steel portal frames at Westway, Section 6, London.

CHAPTER FIVE

Applications

Some of the factors that have been mentioned in this volume have an impact on the design of the whole structure, as illustrated by the following projects.

5.1 Commonwealth Avenue Bridge, Canberra[38]

This bridge comprises two separate multi-cell box-girders continuous over five spans, the main span being 73 m (240 ft) and the total length 311 m (1020 ft). Two Kreutz hardened steel roller bearings were located at the top of each column and at the abutments. In taking into account an allowance for possible foundation settlement and tilt, it was found that, under total concentric load conditions, bearings of 1380 tonnes were required, but each bearing was specified for 1630 tonnes under dead load only to allow for the possible tilt under settlement of the substructure.

The bridge was fixed at one end and a comb joint provided at the other, but either end could be arranged to provide anchorage if required. It was found that rolling friction was only 0·2% in pushing the whole bridge along to the required optimum position. This was well below the design value of 3%. It indicates the high quality of construction rather than that the design friction value can be safely reduced for other projects.

5.2 Huntley's Point Overpass, Sydney[39]

This relatively small overpass illustrates techniques dealing with movements in a curving asymmetrical viaduct. The bridge was anchored at one main pier on two Kreutz rocker bearings. The curvature in plan of the bridge and the assymetrical arrangement of the type of propped cantilever beam deck led to a special system of eccentricity of the loadings, both dead and live, in the structure. These were established by model-testing work and confirmed by the calculations. At the main pier, the bearing on the inside of the curve was specified to be 1775 tonnes capacity and the bearing on the outside of the curve was 760 tonnes capacity. At one end, a tie-down of prestressed Macalloy bars was provided and a double concrete hinge which allowed for shortening of the deck from prestressing and creep without subsequent vertical deflection under live load. On the other side of the main anchorage, rubber bearings were provided at each pier, with differing capacities to allow for the required movements. At the far abutment end, instead of tilted rubber bearings to fix the end span at the abutment, horizontal bearings took the vertical load and two small rubber bearings were interposed between the end block vertically thrusting on the abutment walls.

5.3 Westway (Section 5), London[40]

This continuous curving viaduct has 17 spans of 62·2 m (204 ft) and it was convenient to anchor the whole structure at one end at an anchorage abutment capable of resisting friction forces totalling ±4260 tonnes. Two Glacier sliding bearings are provided, up to 2810 tonnes working capacity at each column, providing up to 203 tonnes lateral correction forces to guide the bridge to the correct alignment. Total movement is up to 825 mm.

5.4 Westway (Section 1), London[41]

The principle of monolithic construction developed on the span-by-span basis has been applied to a complete complex some 1280 m (4200 ft) long involving an elevated roundabout, slip roads and an overpassing structure. The point of stagnant movement is located near the centre of the roundabout and the bearings of the roundabout are arranged to provide free expansion in a radial direction while retaining guides with a gap of ±6 mm ($\frac{1}{4}$ in.) in the tangential direction. Hence, any out-of-balance force is resisted by the sum of the components of the appropriate bearings. The overpass forms a stabilizing strut to the roundabout, taking 80% of the forces which would tend to distort the ring. Bearings are of the Andre Rotaflon type with a capacity ranging from 600 to 1000 tonnes, some with provision for resisting a maximum horizontal force up to 100 tonnes.

5.5 West Gate Bridge, Melbourne

The 67 m (220 ft) approach spans on each side of the main navigation spans consist of a prestressed concrete box girder. The columns are up to about 43 m (140 ft) high and the approaches are horizontally curved. Steel rocker bearings of the Westwood Hi-load type are placed both at the foot of the column and at the top of the column between it and the deck, forming large rollers equal in height to the height of the column. Anchorage is provided at each lower end and the deflection of the columns is related to the movement generated. With the substantial horizontal curvatures involved, the flexibility of the deck and substructure in both the horizontal, vertical and longitudinal directions was arranged to be compatible so that the bridge could be simply pulled back into the correct alignment without undue stressing of any of the elements. Major Demag expansion joints are positioned at the upper ends of each approach.

5.6 Hammersmith Flyover, London[42,43]

This continuous structure employed roller bearings of the Kreutz type placed at the foot of the columns below ground level. The structure was arranged to move on these like a frame on wheels. Each half was anchored at the approach abutments and there is a comb expansion joint approximately at the centre of the flyover. Each approach was substantially straight but the roller bearings were aligned transverse to the exact point of anchorage of each half of the structure.

5.7 Westway (Section 6), London

Bearings at the foot of the columns have also been adopted on a section of the Westway. Metalastik sliding bearings provided at the foot of steel transverse bents range up to 14 MN (1400 tonf) working load to allow movement in the longitudinal direction while providing a restraining force ranging up to 2 MN (200 tonf) laterally so that the portal frame action is achieved. Maximum sliding movement allowed for is 230 mm (9 in.).

5.8 Roll-on, roll-off vehicle ramp terminals

An effective application of rubber and ptfe has been made in the design of several Ro-Ro terminals in Australia. The main girders of the vehicle ramp are supported by a transverse lifting beam which remains in the same vertical plane. The sliding motion takes place during raising or lowering of the ramp and the unusual feature is that, when sliding takes place, the bearings are under very light pressures of approximately 1·4 to 7 N/mm^2 (200 to 1000 lbf/in^2) whereas the short-duration live loading produces a service pressure on the ptfe of up to nearly 21 N/mm^2. This is because the dead load is very small compared to a live load which caters for up to 50 tonne axles plus 30% impact.

In the light of the unusual requirements of these bearings, some tests were carried out on the coefficient of friction particularly under light loads. The coefficient of friction to unstick the bearing, and thereafter at steady sliding between the ptfe and stainless steel interface, are shown for the three bearing pressures involved under dead load.

Coefficient of friction (%)	Vertical bearing pressure N/mm^2 (lbf/in^2) 1·61 (234)	4·47 (648)	7·34 (1065)
to unstick	5·5	4·4	3·5
at steady sliding	3·7	2·5	2·1

The bearings were also tested under a vertical load of 1·78 times the maximum load without distress.

The satisfactory performance of this bearing under test and subsequently in service indicates that there is a wide field of application for rubber and ptfe in moving structures of all types.

5.9 Gladesville Bridge, Sydney[44]

This, the largest clear span in concrete yet to be constructed, is a good example of the use of the flexibility of the structure to accommodate movements. It is a fixed arch, and deck expansion joints are located at the mid-point between abutment and crown; they are thus approximately over the springing of the arch.

5.10 Mancunian Way[18]

The Macunian Way has been referred to in the discussion of the control of lateral forces.

5.11 Narrows Bridge, Perth[45]

This is an example of a deck structure made continuous for live load, anchored at one end and free to move at the other.

5.12 Tasman Bridge, Hobart[46]

The tall columns of this structure have already been discussed.

Conclusion

Despite modern progress there is still a vast future potential for the development of new types of bearings and expansion joints for bridges of all kinds. Bridge engineers will welcome these advances but continue to be eternally wary of doubtful products and design procedures.

top: Commonwealth Avenue Bridge during construction, showing roller bearings at top of piers.

above: Completed Commonwealth Avenue Bridge.

right: Kreutz roller bearings on piers of Commonwealth Avenue Bridge.

above: View showing trimming of bearing pits in Hammersmith Flyover.

left: Hammersmith Flyover during construction. Inserting jacking plate under Kreutz roller bearings in bearing pit at foot of column.

below: Westway, Section 5, London.

right: Metalastik slide bearing in inverted position showing main components. The cylinder plate receives the rubber pad and the piston faced with ptfe material fits into the channel-shape slider plane.

above: Metalastik sliding bearing under test in the inverted position. The compression load is 1400 tonnes and the simulated side load is 300 tonnes in this test on a bearing for Westway, Section 6, London.

right: Metalastik sliding bearing at foot of steel portal frame at Westway, Section 6, London. Protective shields have not yet been bolted into position to the faces of the portal leg.

above: Westway, Section 1, London. The elevated complex near White City.

below: Gladesville Bridge, Sydney, Australia.

REFERENCES

1. WILSON, W. S. Advances in construction methods and equipment. *Journal of the Institution of Civil Engineers. Proceedings*. Vol. 13, February 1940, pp. 273–304.
2. FORRESTER, K. and CURTIS, T. J. Interim report on Gladesville Bridge arch deflections. Institution of Engineers, Australia. *Civil Engineering Transactions*. Vol. C.E. 9, No. 2, October 1967.
3. BRITISH STANDARDS INSTITUTION. BS 153: Part 3A: 1954. *Specification for steel girder bridges. Part 3A. Loads*. London, pp. 59.
4. MINISTRY OF TRANSPORT. *Expansion joints for use in highway bridge decks*. London, 1967, Technical Memorandum (Bridges) No. BE6.
5. CAPPS, M. W. R. *The thermal behaviour of the Beachley Viaduct/Wye Bridge*. Crowthorne, Road Research Laboratory, 1968, pp. 46, RRL Report LR 234.
6. EMERSON, M. *Bridge temperatures and movements in the British Isles*. Crowthorne, Road Research Laboratory, 1968, pp. 38, RRL Report LR 228.
7. CAPPS, M. W. R. *Temperature movements in the Medway Bridge*. Interim report. Harmondsworth, Road Research Laboratory, 1965, Laboratory Note No. LN/914/MWRC. Unpublished.
8. EMERSON, M. *Temperature movements in the Hammersmith Flyover*. Harmondsworth, Road Research Laboratory, 1966, Technical Note No. 33. Unpublished.
9. CAPPS, M. W. R. *Thermal bending of the river span of the Medway Bridge*. Harmondsworth, Road Research Laboratory, 1965, Technical Note No. 20. Unpublished.
10. BONNELL, D. G. R. and HARPER, F. C. *The thermal expansion of concrete*. London, HMSO, 1951, pp. 21, National Building Studies Technical Paper No. 7.
11. STEPHENSON, D. A. Effects of differential temperature on tall slender columns. *Concrete and Constructional Engineering*. Vol. 56, No. 5, May 1961, pp. 175–178; No. 11, November 1961, pp. 401–403.
12. LEONHARDT, F. and ANDRÄ, W. Stutzungsprobleme der Hochstrassenbrücken. *Beton- und Stahlbetonbau*. Vol. 55, No. 6, June 1960, pp. 121–132. English translation: London, Cement and Concrete Association, April 1962, pp. 30. Library Translation Cj. 95.
13. ROARK, R. J. *Formulas for stress and strain*. Fourth edition. New York, McGraw-Hill Book Company, 1965, p. 157.
14. O'DONNELL, W. J. The additional deflection of a cantilever due to the elasticity of the support. *Transactions of the American Society of Mechanical Engineers. Journal of Applied Mechanics*. Vol. 27, No. 3, September 1960, pp. 461–464.
15. FAUPEL, J. H. *Engineering design*. New York, John Wiley & Sons Inc., 1964.
16. MULLER, J. Concrete bridges built in cantilever. *Journal of the Institution of Highway Engineers*. Vol. 10, No. 4, October 1963, pp. 156–171; Vol. 11, No. 1, January 1964, pp. 19–30.
17. ZEDERBAUM, J. The frame action of a bridge deck supported on elastic bearings. *Civil Engineering & Public Works Review*. Vol. 61, No. 714, January 1966, pp. 67–72.
18. BINGHAM, T. G. and LEE, D. J. The Mancunian Way elevated road structure. *Proceedings of the Institution of Civil Engineers*. Vol. 42, April 1969, pp. 459–492.
19. PRICE, W. I. J. and STOTT, J. P. *Provision and performance of joints and bearings in roads and bridges*. Paper presented to the symposium on 'Methods of accommodating movement in prestressed concrete structures', London, November 1964. London, Prestressed Concrete Development Group, November 1964, pp. 20.
20. KÖSTER, W. *Expansion joints in bridges and concrete roads*. Second edition. London, Maclaren & Sons, 1969, pp. 333.
21. WATSON, S. C. *Joint sealing practice for longer spans*. American Association of State Highway Officials, 1967.
22. WATSON, S. C. *New developments in joint sealing practice for longer spans*. American Association of State Highway Officials, 1968.
23. WATSON, S. C. *A concept of pre-engineered, prefabricated, prestressed modular and multi-modular sealing systems for our modern bridges and structures*. Paper presented to the Symposium on the Engineering of Bearings and Expansion Joints for Bridges, Johannesburg, 11–13 June 1969.
24. SIMS, F. A. and BRIDLE, R. J. The design of concrete hinges. *Concrete and Constructional Engineering*. Vol. 59, No. 8, August 1964, pp. 277–286.
25. BASE, G. D. *Tests on four prototype reinforced concrete hinges*. London, Cement and Concrete Association, May 1965, pp. 28, Research Report No. 17.
26. BASE, G. D. *The development of reinforced concrete hinges of large load and rotation capacity*. London, Cement and Concrete Association. To be published as a Research Report.
27. MINISTRY OF TRANSPORT. *Rules for the design and use of Freyssinet concrete hinges in highway structures*. London, HMSO, 1966, pp. 14, Memorandum No. 577/1.
28. TAYLOR, M. *Rubber bridge bearings. A survey of present knowledge*. Harmondsworth, Road Research Laboratory, April 1965. pp. 22, Laboratory Note No. LN/822/MET.
29. LINDLEY, P. B. *Engineering design with natural rubber*. London, Natural Rubber Producers Research Association, 1964, Technical Bulletin No. 8.
30. MINISTRY OF TRANSPORT. *Provisional rules for the use of rubber bearings in highway bridges*. London, HMSO, 1962 (corrections January 1963), pp. 19, Memorandum No. 802.
31. INTERNATIONAL UNION OF RAILWAYS. *Code for use of rubber bearings for rail bridges*. UIC Code 772R. Paris, January 1969, pp. 56.
32. BRITISH STANDARDS INSTITUTION. BS 1154: 1952. *Vulcanized rubber compounds*. London, pp. 8.
33. BRITISH STANDARDS INSTITUTION. BS 903. *Methods of testing vulcanized rubber*. Part A2: 1956. *Determination of tensile stress–strain properties*, London, pp. 16; Part A6: 1963, *Determination of compression set*, pp. 16; Part A7: 1957, *Determination of hardness*, pp. 12; Part A19: 1956, *Accelerated ageing tests*, pp. 16.
34. VETZ, H. and HAKENJOS, V. Gleitreibungsuntersuchungen mit Polytetrafluoräthylen bei hin- und heigehender

Bewegung. *Die Bautechnik*. Vol. 10, No. 5, May 1967, pp. 159–166.
35. LONG, J. E. The performance of p.t.f.e. in bridge bearings. *Civil Engineering and Public Works Review*. Vol. 64, No. 754, May 1969, pp. 459–462.
36. LEONHARDT, F., BAUR, W. and TRAH, W. Brucke über den Rio Caroni, Venezuela. *Beton- und Stahlbetonbau*. Vol. 61, No. 2, February 1966, pp. 25–38.
37. BOBISCH, G. *Practical difficulties and problems regarding bearings and expansion joints in bridges*. Paper presented to the Symposium on the Engineering of Bearings and Expansion Joints for Bridges, Johannesburg, 11–13 June 1969.
38. BIRKETT, E. M. and FERNIE, G. N. Bridges in the Canberra Central Lake Area—Design. *Journal of the Institution of Engineers (Australia)*. Vol. 36, No. 7–8, July–August 1964, pp. 000–000.
39. LEE, D. J. Huntley's Point Overpass. *Structural Concrete*. Vol. 2, No. 12, November/December 1965, pp. 521–534.
40. LEE, D. J. Western Avenue Extension—The design of section five. *The Structural Engineer*. Vol. 48, No. 3, March 1970, pp. 109–120.
41. LEE, D. J. *Prestressed concrete elevated roads in Britain*. London, Concrete Society, 1967. pp. 27. Technical Paper PCS 12. Also published in *Journal of the Prestressed Concrete Institute*. Vol. 13, No. 6, December 1968, pp. 67–89.
42. RAWLINSON, Sir JOSEPH and STOTT, P. F. Hammersmith Flyover. *Journal of the Institution of Civil Engineers. Proceedings* Vol. 23, December 1962, pp. 565–600.
43. WROTH, C. P. The Hammersmith Flyover—site measurements of prestressing losses and temperature movement. *Journal of the Institution of Civil Engineers. Proceedings* Vol. 23, December 1962, pp. 601–624.
44. BAXTER, J. W., GEE, A. F. and JAMES, H. B. Gladesville Bridge. *Journal of the Institution of Civil Engineers. Proceedings* Vol. 30, March 1965, pp. 489–530; Discussion Vol. 34, June 1966, pp. 244–257.
45. BAXTER, J. W., BIRKETT, E. M. and GIFFORD, E. H. The Narrows Bridge, Perth, Western Australia. *Journal of the Institution of Civil Engineers. Proceedings* Vol. 20, September 1961, pp. 39–84; Discussion Vol. 22, May 1962, pp. 113–126.
46. NEW, D. H., LOWE, J. R. and READ, J. The superstructure of Tasman Bridge, Hobart. *The Structural Engineer*. Vol. 45, No. 2, February 1967, pp. 81–90.

APPENDIX 1: Extracts from various standards relating to temperature effects

United Kingdom

BS 153: Part 3A: 1954

'Allowances shall be made for the forces resulting from the following conditions:

(a) Any portion of the superstructure being restrained from moving when subjected to variations in temperature.

For this purpose in Great Britain a minimum of 20°F and a maximum between 80°F and 120°F, depending on the location of the structure, shall be taken. Elsewhere the temperature limits shall be based on local conditions.

(b) Any portion of the superstructure being at a temperature different from the rest of the structure, due to the effect of sun and shade. For this purpose the maximum difference of temperature shall be taken as 15°F.

In determining forces and movements due to change of temperature the coefficient of expansion of steel shall be taken as 0·0000065 per Fahrenheit degree.'

(Note: 20°F = −7°C; 80°F = 27°C; 120°F = 49°C; 15 degF = 8 degC; 0·0000065 per °F = 0·0000117 per °C.)

Ministry of Transport Technical Memorandum on Expansion Joints for Use in Highway Bridge Decks. No. BE 6

Clause 6 (c), (ii):

'(ii) *Total thermal movement*
The total range of thermal movement that may be expected in Great Britain can then be calculated from the figures given in the table below.

Type of construction	Extremes of mean bridge temperature: °C or °F	High	Low	Total range of thermal movement (inches/100 ft)
Steel box girders with steel deck	°C (°F)	49 (120)	−7 (20)	0·80
Composite construction	°C (°F)	43 (110)	−7 (20)	0·72
Concrete box or I-beam deck	°C (°F)	38 (100)	−7 (20)	0·64
Solid slab deck	°C (°F)	32 (90)	−7 (20)	0·56

Notes
1. The range of movement given in the last column is based on an average coefficient of expansion of 0·000012 per °C (0·0000067 per °F) for all types of deck.
2. The mean temperature of a bridge deck is the average temperature of the structure taking account of the temperature gradient occurring through the structure.'

USA

The following extracts are quoted from the Ninth Edition (1965) of the AASHO bridge specifications:

'Section 1. General features of design
1.2.15 Thermal forces
Provision shall be made for stresses or movements resulting from variations in temperature. The rise and fall in temperature shall be fixed for the locality in which the structure is to be constructed and shall be figured from an assumed temperature at the time of erection. Due consideration shall be given to the lag between air temperature and the interior temperature of massive concrete members or structures.

The range of temperature shall generally be as follows:
Metal structures
Moderate climate, from 0° to 120°F.
Cold climate, from −30° to 120°F.
Concrete structures
Moderate climate, temperature rise 30°F, temperature fall 40°F
Cold climate, temperature rise 35°F, temperature fall 45°F.

Section 5. Concrete design
In general, provision for temperature changes shall be made in all simple spans having a clear length in excess of 40 feet.

In continuous bridges, provision shall be made in the design to resist thermal stresses induced or means shall be provided for movement caused by temperature changes.

Expansion not otherwise provided for shall be provided by means of hinged columns, rockers, sliding plates or other devices.

Section 7. Structural steel design
1.7.17 Expansion and contraction
The design shall be such as to allow for total thermal movement at the rate of $1\frac{1}{4}$ in. in 100 feet. Provisions shall be made for changes in length of span resulting from live load stresses. In spans more than 300 feet long, allowance shall be made for expansion and contraction in the floor. The expansion end shall be secured against lateral movement.'

Further information

The coefficient of thermal expansion for concrete structures is assumed to be 0·000006 per degF and for structural steel 0·0000065 per degF.

The American Railway Engineering Association provides for thermal expansion in steel structures at the rate of 1 in. in 100 ft (1 in 1200) and in concrete structures at a computed rate using a coefficient of thermal expansion of 0·0000065 per degF and a temperature range determined by the individual railroad.

It should be noted that although the *Standard Specifications for Highway Bridges* are the accepted standard specifications for the design of highway bridges in the USA the various state highway departments modify these specifications to allow for local conditions.

Germany

The following are translated extracts from DIN 1072, *Road and Foot Bridges: Design Loads*, November 1967.

6.1.1 Thermal effects on steel bridges
A temperature range of ± 35 degC should be allowed for about an average temperature of 10°C.

On walled constructions, or on the surfaces of walls above ground, when the place where the variation of temperature occurring is known, a linear variation of 15 degC should be allowed for. [See Explanations, DIN 1072 (supplement).]

Between different construction elements, a temperature difference of 15 degC should be allowed for, i.e. between tie-rod and arch, between cables and stiffeners, between top and bottom flanges of frameworks, etc.

If, in a multiple girder bridge, the sunlight acts on the edge girder, then a variation of only 5 degC should be taken for all the remaining girders. If no effective pattern of sunlight is given, it is not necessary to consider a temperature differential for a single girder.

The effects due to temperature fluctuation and temperature range in both vertical and horizontal directions should be added together. In the case of steel bridges with reinforced concrete slabs acting as the deck, the clauses for composite section bridges apply. [See Section 6.1.2 and Explanations, DIN 1072 (supplement).]

6.1.2 Thermal effects on composite bridges
To take account of the thermal effects two extremes should be considered about a temperature level of $+10$°C.

(a) Temperature increase at the top of the concrete slab of about 20 degC and at the bottom flange of the steel girder of about 35 degC.

(b) Temperature decrease at the top of the concrete slab of about 20 degC and at the bottom flange of the steel girder of about 35 degC.

The temperature variation should be taken as a linear variation for (a) or (b). [See Explanation, DIN 1072 (supplement).]

Temperature variation for single member should be taken in accordance with clause 6.1.1.

6.1.3 Thermal effects on massive bridges
A temperature range of ± 20 degC should be allowed for about an average temperature of $+10$°C.

For structural members, the smallest dimension should be at least 700 mm; the same applies to the thickness of a layer of concrete being poured, or steps should be taken to ensure that the temperature variation is kept within 5 degC.

In ascertaining the smallest dimension, it should be borne in mind that the presence of hollow voids in the concrete should be avoided (i.e. in box-sections), unless their cross-section does not amount to 50% of the total cross-section of the member concerned.

Uneven heating of individual members need rarely be taken into account if considered within the temperature range. (For instance between tie-rod and arch on two-hinged arches.) If necessary it should be taken as ± 5 degC. [See Explanations, DIN 1072 (supplement).]

6.1.4 Thermal effects on timber bridges
Temperature variations can be permanently ignored.

Additional considerations
To Section 6.1.3
Because of the small magnitude of the stresses caused by the hardening of the concrete which originate when deep sections are dealt with, it might be well worth while to assume a larger temperature range, so that the above-mentioned stresses are taken care of. [See Explanations to Section 6.1.3, DIN 1072 (supplement).]

Explanations
To Section 6.1: Thermal effects
The individual sections deal more thoroughly with the problems of the three types of construction described than was done in previous editions. The temperature variation adopted seems to have been satisfactory in practice.

To Section 6.1.1: Thermal effects on steel bridges
Different temperatures in the flanges and walls due to the influence of sunlight can be large. This is not always a governing condition and sometimes it can have a neglectable influence. On orthotropic decks, for instance, the slab acts as a radius for the heat whereas the web can induce higher or lower temperatures over the flange depending on its depth.

For steel bridges with concrete slabs, the thermal effects for composite construction can be applied, even though no shear force is considered between slab and girder.

To Section 6.1.2: Thermal effects on composite bridges
The statements on temperature variation made in DIN 1078 are not quite realistic and should be changed.

To Section 6.1.3: Thermal effects on massive bridges
The uneven heating of individual members of ± 5°C (for instance between tie-rod and arch) should be expected when both concrete and reinforced concrete exist. On massive arches with free horizontal tie-rods or on prestressed bridges with inclined cables uncovered by concrete or any other combination of steel and concrete members the considerations given in Section 6.1.2 should apply as a whole.

For these clauses, the aim has been to prevent the concrete from cracking due to stress caused by thermal effects.

Measures which can relieve these effects are, for instance, the cooling of aggregates, cement and water, the use of a low-heat cement, the sub-division of concreting into more sections, or even the installation of refrigerating ducts inside the concrete mass.

APPENDIX 2: Code for the use of rubber bearings for rail bridges

UIC Code 772 R: Published by the International Union of Railways and reproduced by permission. First edition 1 January 1969.

Introduction

This code is based on studies carried out from 1961 to 1965 by the D 60 Committee of Experts of the Office for Research and Tests (ORE), at the request of the UIC 7th Committee. These studies resulted in the drawing up of 'Recommendations' (Appendix 2 of the Final ORE Report D 60/RF dated October 1965). The many tests carried out for this report allowed formulae to be proposed for the design of the bearings, which were prepared in relation to the test results and are dealt with more specifically in Section 3 below.

Attention is drawn to the fact that the text of the directives and that of Appendix 2 'Design example' make use of the *bar* as a unit of stress and the *decanewton* as a unit of force. One bar is equivalent to 1·02 kgf/cm^2 and one decanewton 1·02 kgf.

Section 1: Purpose of the code

The purpose of the code is to define the scope for the use of rubber bearings, the types to be used, the design of the bearings, their placing, and also the precautions to be taken with regard to inspection and maintenance.

The only bearings to which the code is intended to apply are the two kinds described below, to the exclusion of all other bearings of various shapes and characteristics (1).

1.1 Unreinforced bearings
Unreinforced bearings consist of only one rubber layer in the form of a rectangular block, usually elongated.

1.2 Reinforced bearings

Reinforced bearings consist of a certain number of layers of rubber of rectangular shape separated by plates, also rectangular, of sheet steel with or without perforations, which, during manufacture, are strongly bonded to the layers of rubber to stiffen them. The sides of the steel plates may be visible or encased in rubber during manufacture. The same applies to the external faces of the top and bottom steel plates. When the bearing is completely encased, it has the appearance of a uniform block.

OBSERVATIONS

(1) Bearings not referred to in the code will be the subject of special regulations.

Section 2: Properties and use

2.1 Properties

The bearings can withstand heavy vertical loads while allowing considerable lateral movement and rotation.

2.2 Use

They are used as movable bearings (1), the flexibility of the rubber allowing both rotation and tangential movement of the structure (2) supported within the limits laid down below.

Correct functioning, however, requires a sufficient load to avoid creep, and a sufficiently high deflection under vertical load, to ensure that the rotation of the structure does not bring about total unloading of a section of the bearing surface. However, this elasticity must not be so large as to produce secondary forces in the structure carried (3).

2.2.1 Unreinforced bearings

By reason of their flexibility, these are used for distributed loads of moderate intensity, such as those emanating from concrete slab bridges or joist-in-concrete bridges less than 20 m long between expansion joints.

2.2.2 Reinforced bearings

These are used for heavy concentrated loads such as those produced by heavily loaded beams.

2.2.3 Effect of the deflection of the bearings on the structure they support

The deflection of rubber bearings has little effect on the eccentricity of vertical loads, even when rotation occurs.

Conversely, these bearings do not distribute the loads uniformly over the bearing surface; pressure at the edge is always less and in the centre it is 1·5 to 2 times the average pressure. The design of the reinforcement of the bedstones and the bearing stiffeners of the structure carried should take this into account.

2.2.3.1 Simply supported bridges

The deck of these structures, which usually includes a line of movable bearings, can be calculated by neglecting the stresses due to differential vertical deflections of the line of bearings. Conversely, horizontal deflections are accompanied by forces, the transmission of which should be examined both in the substructure and the superstructure.

2.2.3.2 Continuous bridges

In principle, these have only one line of fixed bearings. The vertical deflections of the movable rubber bearings enter only rarely into the calculation of the deck, the compression of these bearings usually being negligible in relation to that of the supports. Horizontal deflections should be examined in the same way as for simply supported spans, particularly as far as checking the stability of intermediate piers is concerned.

OBSERVATIONS

(1) Generally speaking, rubber bearings of the kinds referred to above are not usually used as fixed bearings. The use of a fixed bearing could only be allowed after special studies and tests.

(2) Tangential movement causes appreciable horizontal forces which must be taken into account in the design of the supports, especially when these supports are in the form of piers or columns. These horizontal forces also act on the fixed bearings. Unlike metal expansion bearings, horizontal reactions of rubber bearings are proportional to the movement.

(3) Rubber bearings are not recommended in cases where they would be subject to small loads and considerable longitudinal movements at the same time. This would be the case for example, with bearings intended for metal rail bridges with a small span, or in certain cases of end spans of continuous bridges.

Section 3: Design (1)

It is necessary to distinguish between prolonged and rapidly changing loads (2). The latter have a much more severe effect on the rubber because of their frequent occurrence.

3.1 Plan dimensions

The dimensions a and b, parallel and at right-angles respectively to the beam, are related to those of the beams and bedstones; in particular, b must generally be much the same as the width of the beam.

In addition, the area $A = ab$ must be such that the average pressure (3) remains between two limits corresponding respectively to the maximum permissible pressure on the bearing and on the bedstone, and a minimum corresponding to the no-slip condition.

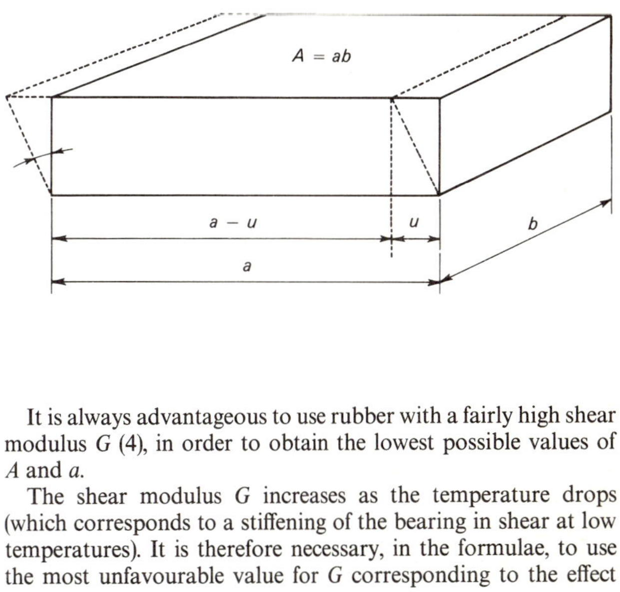

It is always advantageous to use rubber with a fairly high shear modulus G (4), in order to obtain the lowest possible values of A and a.

The shear modulus G increases as the temperature drops (which corresponds to a stiffening of the bearing in shear at low temperatures). It is therefore necessary, in the formulae, to use the most unfavourable value for G corresponding to the effect under consideration.

In calculating the bearing area, the external casing of the reinforcement is not taken into account.

3.1.1 Unreinforced bearings

3.1.1.1 The mean calculation pressure must satisfy the condition (5):

$$\sigma_m < 2GS \text{ with } S = \frac{ab}{2T(a+b)}$$

3.1.1.2 No slip will be ensured by the two conditions:

1. $$\sigma'_m = \frac{P_c}{A'} > 10\left(1 + \frac{a}{b}\right) \text{ bars*} \qquad (6)$$

* 1 bar equals 1 kgf/cm² to within 2%.

2. The most unfavourable of the two inequalities (7):

$$H_c < f'P_c$$

or

$$H_c + H_s < f(P_c + P_s)$$

with:

H_c = slowly applied tangential load
P_c = slowly applied normal loads
H_s = quickly applied tangential load
P_s = quickly applied normal loads
f or f' = coefficient of friction

3.1.2 Reinforced bearings

3.1.2.1 It is not sufficient to consider normal loads (8) *only when determining the permissible load for these bearings. It is also necessary to consider tangential loads which depend on the thickness of the bearings.*

3.1.2.2 The no slip-conditions are (9):

1. $$\sigma'_m = \frac{P_c}{A'} > 20 \text{ bars}$$

2. The most unfavourable of the two inequalities:

$$H_c < f'P_c$$

or

$$H_c + H_s < f(P_c + P_s)$$

3.2 Thickness (10)

This dimension is governed by the need to limit either the internal deformation of the rubbber or the stresses in it.

3.2.1 Unreinforced bearings

3.2.1.1 Limiting of shear movement (11)

$$\tan \gamma_{c,s} < 0{\cdot}70, \qquad T > 1{\cdot}43\,u$$

3.2.1.2 Condition requiring no uplift of the most lightly loaded edge of the bearing (12)

The tangents of the angles of rotations α_c and α_s imposed on the bearing must satisfy the more unfavourable of the following two conditions:

1. $$\tan \alpha_c < \frac{2e_c}{a}$$

with

$$e_c = \frac{T\sigma'_m}{10GS + 2\sigma'_m} \quad \text{and} \quad \sigma'_m = \frac{P_c}{A'}$$

2. $$\tan \alpha_c + 1{\cdot}5 \tan \alpha_s < \frac{2e}{a} \qquad (13)$$

with

$$e = \frac{T\sigma_m}{10GS + 2\sigma_m} \quad \text{and} \quad \sigma_m = \frac{P_c + P_s}{A'}$$

3.2.1.3 Stability condition of the bearing

$$T < \frac{a}{5}$$

3.2.2 Reinforced bearings (14)

For these bearings, it is also necessary to limit the shear stresses to avoid separation of the rubber and the steel plates.

3.2.2.1 Limiting of shear movement (15)

$$\tan \gamma_{c,s} < 0{\cdot}70, \qquad T > 1{\cdot}43\,u$$

3.2.2.2 Condition requiring no uplift of the most lightly loaded edge of the bearing (16)

Tan α_c and tan α_s must satisfy the more unfavourable of the following conditions:

1. $$\tan \alpha_c < \frac{6\Sigma e'_i}{a}$$

with

$$e'_i = \frac{t_i\sigma'_m}{4GS_i^2 + 3\sigma'_m}$$

calculated deflection of the index layer i under the vertical load:

$$\sigma'_m = \frac{P_c}{A'} \quad \text{and} \quad S_i = \frac{ab}{2t_i(a + b)}$$

2. $$\tan \alpha_c + 1{\cdot}5 \tan \alpha_s = \frac{6\Sigma e_i}{a}$$

with

$$e_i = \frac{t_i\sigma_m}{4GS_i^2 + 3\sigma_m}$$

calculated deflection of the index layer i under the vertical load:

$$\sigma_m = \frac{P_c + P_s}{A'} \quad \text{and} \quad S_i = \frac{ab}{2t_i(a + b)}$$

3.2.2.3 Stability condition for the bearing (17)

$$\sigma_m < \frac{2a}{3T} GS_i$$

where S_i has the value given in 3.2.2.2, calculated for the thickest layer.

3.2.2.4 Limiting of the shear stress in the rubber (18)

The stress τ should satisfy the following condition:

$$\tau = \tau_p + \tau_H + \tau_\alpha < 5G$$

In view of the fact that u and α can have different signs for slowly applied loads and quickly applied loads, it is also necessary to check the condition under permanent load only (19).

With fully encased bearings, the limit imposed to τ can be raised if this is justified by the result of the tests carried out on acceptance of bearings of the same type.

3.2.2.5 Strength of the steel plates

The steel plates are subjected to tensile stresses resulting from the vertical load.

The thickness of the steel plate (20) situated between two layers of rubber of respective thicknesses t_1 and t_2 should comply with the conditions:

$$t_s > \frac{2(t_1 + t_2)(P_c + 1{\cdot}5\,P_s)}{A'\sigma'_a}$$

where σ'_a is the permissible tensile stress (21) for the steel used.

OBSERVATIONS

(1) For design, a number of conditions has to be checked.

The order followed in the code and the design example in its appendix enable the trial and error process to be limited and the various possible designs to be obtained quickly.

(2) Loads of long duration (dead loads, loads due to temperature or shrinkage) will have the index c applied to them. Quickly applied loads of rapid variation (live loads) will have the index s applied to them.

(3) The average pressure is defined by:

$$\sigma_m = \frac{P}{A'}$$

with

$$P = P_c + P_s$$

and

$$A' = (a - u)b$$

in which u is the horizontal shear strain.

(4) *The shear modulus G is defined by the quotient:*

$$G = \frac{\tau}{\tan \gamma}$$

where τ is the shear stress H/A of the rubber under a tangential load H, and $\tan \gamma$ is the strain u/T of the rubber under the same load H, with T = total thickness of the rubber, excluding steel plates.

Therefore, in addition:

$$G = \frac{H}{A} \cdot \frac{T}{u}$$

In practice, the value of the modulus G is related to the duration of the load, and it is sufficient to consider two values of G depending on whether the load is applied slowly or quickly. The 'dynamic' modulus G_s is more or less double the 'static' modulus G measured under prolonged loads:

$$G_s \doteq 2G$$

(5) *This condition is equivalent to limiting the vertical deflection of the bearing to 15% of its thickness. The vertical deflection of a pad is actually equal to:*

$$e = \frac{T\sigma_m}{10GS + 2\sigma_m}$$

(6) *Tests have shown that f' was at least*

$$0{\cdot}10 + \frac{2}{\sigma'_m}$$

where

$$\sigma'_m = \frac{P_c}{A'}$$

in bars (kgf/cm^2)

For quickly applied vertical loads acting at the same time, the value of f may become

$$0{\cdot}10 + \frac{2}{\sigma_m} \quad with \quad \sigma_m = \frac{P_c + P_s}{A'}$$

Higher values can be obtained by bonding for example, but these must be proved by test if they are to be used in design.

(7)
$$H_c = GA\frac{u_c}{T}$$

$$H_s = G_s A\frac{u_s}{T} = 2GA\frac{u_s}{T}$$

(taking into account the fact that the dynamic shear modulus is approximately double the static shear modulus, see observation 4).

(8) *As a guide, bearings at present in service can withstand purely vertical loads of about 60 to 120 bars.*

(9) *The same notation and observations apply as in (6).*

(10) *When the rubber casing (above or below the bearing) is less than half as thick as a single pad and does not exceed 3 mm, it may be considered as a simple protection and need not, therefore, be considered in calculating the deflections.*

If the rubber casing does not fulfil the above conditions, its deflection shall be considered as that of a reinforced pad 1·4 times its actual thickness.

(11) *The deflection under tangential loads H_c (temperature, shrinkage) and H_s (overloads) is equivalent to*

$$\tan \gamma_{cs} = \frac{u}{t} = \frac{H_c}{GA} + \frac{H_s}{2GA}$$

(12) *In particular, $\tan \alpha_c$ must take into account the dead load deflection and placing tolerances.*

(13) *The factor 1·5 by which $\tan \alpha_s$ is multiplied has been introduced to take into account the harmful effect of the varying vertical loads observed during the tests. $\tan \alpha_c$ and $\tan \alpha_s$ may have values of opposite signs.*

(14) *For these bearings, the total height h is:*

$$h = T + \Sigma t_s = \Sigma t_i + \Sigma t_s$$

where T = sum of the single thicknesses t_i of the layers of rubber = total thickness of the rubber, and Σt_s = sum of the thicknesses of the reinforcement.

(15) *See observation 11.*

(16) *See observation 12.*

(17) *This condition is generally satisfied if*

$$T < \frac{a}{5}$$

(18) *This stress τ is the sum of the following shear stresses:*

$$\tau_p = \frac{1{\cdot}5}{S_i}\left(\frac{P_c + 1{\cdot}5P_s}{ab}\right)$$

which result from the vertical load (S_i is defined in 16).

The factor 1·5 by which P_s is multiplied takes into account the harmful effect of the varying vertical loads observed during the tests.

$$\tau_H = G\frac{u}{T} = \frac{H_c + 0{\cdot}5\,H_s}{A}$$

due to the tangential loads (see 11)

$$\tau_\alpha = \frac{Ga^2}{2t_iT}(\tan \alpha_c + 1{\cdot}5 \tan \alpha_s)$$

due to the rotational movement imposed on the bearing (see 12). When the bearing consists of rubber pads of different thicknesses t_i, the calculation is made for the most unfavourable value of τ_α, i.e. by taking the smallest of the thicknesses t_i.

(19) *In this case the shear stresses are expressed as follows:*

$$\tau_p = \frac{1{\cdot}5\,P_c}{S_i ab}$$

$$\tau_H = \frac{Gu_c}{T} = \frac{H_c}{A}$$

$$\tau_\alpha = \frac{Ga^2}{2t_iT}\tan \alpha_c$$

(20) *Or the thickness equivalent to the net section of the steel plate if the latter is perforated.*

(21) *The value σ'_a is fixed by each Administration, taking into account any fatigue effect.*

Section 4: Placing

4.1 Positioning

This should be such that the bearings support both the vertical and horizontal loads transmitted by the deck, at the same time allowing for movement due to live loads, temperature, shrinkage and creep.

If it is necessary to have fixed bearings and rubber bearings (1) along the same line of bearings, the effect on the substructure and superstructure of the different vertical stiffeners should be checked. As an alternative, rubber bearings only may be used, with a separate fixed point.

The arrangement consisting of placing two or more bearings one behind the other along the longitudinal centre-line of the bridge, in the same group of bearings, should not be allowed in any circumstances.

Because of their different stiffnesses, bearings of different sizes must not be placed next to each other.

4.2 Levelling

The contact surface between the bearing and the bedstone must be perfectly prepared and, in principle, horizontal, in order to avoid tangential loads due to gravity.

When the deck is sloping, the slope is usually sufficiently small (a few parts per thousand) to be taken up by the deflection of the bearing itself. If this is not so, it is advisable to provide a horizontal bearing surface, even by varying the thickness of the plate situated between the upper face of the bearing and the lower face of the beam (2).

4.3 Limitation of transverse movements of bridge deck
Since rubber bearings do not resist transverse movement arising either from transverse forces (wind, nosing forces, etc.) or from skew effect, it will often be necessary to provide, between the bearings and the structure carried, a stop device restricting such movement to acceptable limits.

4.4 Anti-creep device
Although resistance to longitudinal slipping of the bearing against the bedstone or the structure is theoretically guaranteed by friction, it is advisable to prevent such slip by means of an anti-creep device. This device should nevertheless allow for longitudinal adjustment of the bearings in certain cases (3).

It is not as worth while, however, near fixed bearings or stops for eliminating tangential movement between the substructure and the superstructure.

4.5 Adjustment
The same precautions are necessary in plan and level as for metal bearings.

The deck and its bearings (after addition of any temporary conditions necessary) must be adjusted to final level so that it rests on wedges or jacks. The bearings must then be packed by suitable means (packing with cement mortar, resin, grouting, bonding adhesive, etc.).

OBSERVATIONS

(1) *This happens in very wide structures, where there must be freedom for lateral expansion of the superstructure.*
(2) *Sections along the longitudinal axis of the structure.*

Reinforced concrete structures
Solution 1 (applicable to end and intermediate bearings)

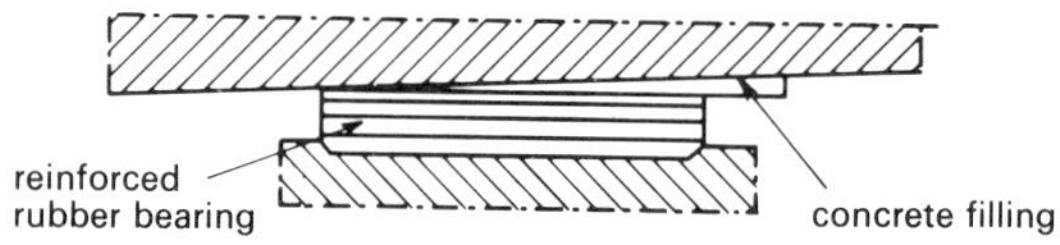

Solution 2 (horizontal intrados on movable bearing)

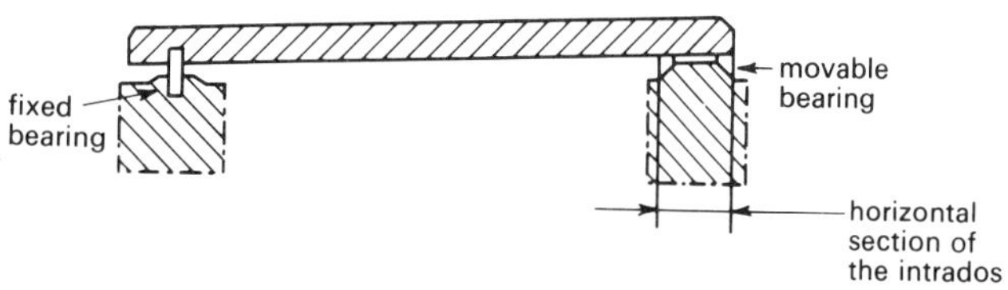

Other structures

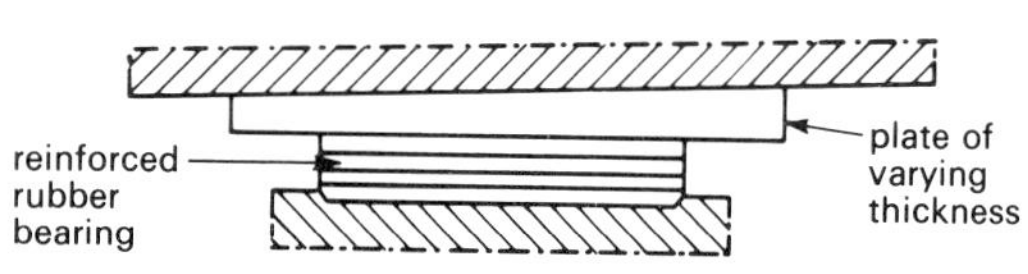

(3) *Particularly in structures where shrinkage or creep are likely to lead to shortening of the beams producing excessive horizontal movement in the bearings (especially bridges in prestressed concrete): there must be provision for the beams to be raised after some months to eliminate this deflection, and to replace the bearings correctly under the beams which have suffered permanent distortion.*

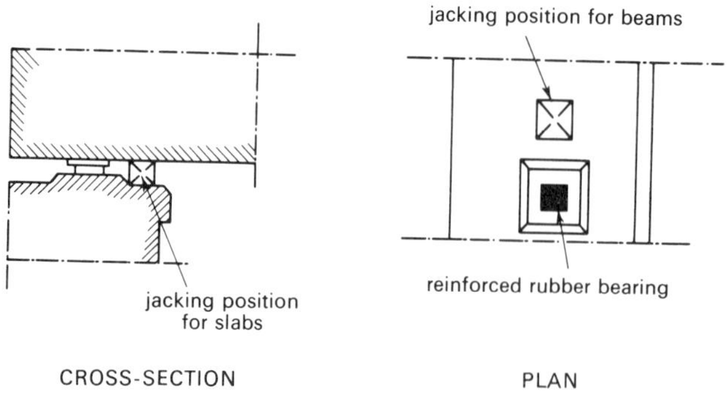

Section 5: Inspection, maintenance and replacement

5.1 Inspection
Inspection shall comprise:

Confirmation that the bearings are satisfactory in service (no creep, no sticking, no abnormal or uneven distortion);

The condition of the bearing after some time: no cracks or deterioration as a result of oxidation (especially with bearings with no outer rubber covering);

The cleanness of the bearings and their immediate surroundings. Particular care must be taken to avoid contact with grease, oil, petrol, etc.

5.2 Maintenance
Special attention will be paid, where necessary, to the painting of the visible parts of the steel plates (special paint).

5.3 Replacement of bearings
This must always be provided for. If it is not possible to use temporary props, jacking pockets must be provided near each bearing to allow for easy removal and replacement.

Appendix 1: Materials, tests and tolerances

Contents

1 General

1.1 Purpose of the Appendix

This Appendix deals with:

the materials employed for movable bearings for railway bridges involving the use of rubber (natural or synthetic);
the tests relating to the acceptance of these bearings;
the tolerances to be applied to the dimensions of the latter.

1.2 Authorization

The types of movable bearing involving the use of rubber must be the subject of prior agreement by the Administration. This agreement will be granted after tests similar to those carried out by the D 60 Committee of Experts of the Office for Research and Tests (ORE) of the International Union of Railways, with additions, where necessary, to meet the proposed characteristics. Particular attention is drawn to the need for carrying out stringent tests at the low temperatures to which the bearings will be subjected.

Attention is drawn to the fact that the minimum tests to be carried out for the purpose of authorization are as follows.

1.2.1 Static tests

Determination of the mechanical properties of the materials used in the various types to be designed (rubber, steel for reinforcement of bearings), also composition of the rubber (rubber, ash, carbon black).

Determination of the mechanical properties of the various samples of bearing (particularly the modulus of elasticity *E* and the shear modulus *G*).

1.2.2 Dynamic tests

Tests on specimens under axial and eccentric vertical load pulsations for constant angles of shear.

Tests on specimens under shear angle pulsations for constant axial compression.

1.2.3 Tests at extreme temperatures

Study of behaviour at low temperatures and other properties of the rubbers (natural or synthetic) used in the manufacture of the bearings submitted for authorization.

Determination of the moduli *E* and *G* of the bearings at extreme temperatures.

Determination of the effect of a period of exposure to cold on the static moduli *E* and *G*.

1.3 Scope

In the same way as the Code, this Appendix is applicable only to types of bearings in the form of a rectangular block comprising one or more layers of natural or synthetic rubber.

Bearings merely consisting of separate bearing pads laid one on top of the other are not allowed.

When the types of bearing or the properties of the materials proposed differ from the types of bearing subjected to the tests of the ORE D 60 Committee of Experts, it may be necessary to envisage conditions of authorization, acceptance and use, other than those given in this Code.

1.4 Acceptance tests

Once the types of bearings supplied to the Administration have been authorized, acceptance of the bearings will consist of conformity tests intended to ensure that the delivery definitely complies with the conditions stipulated by the Administration when the type in question was authorized.

These tests shall be carried out either for the purpose of checking the manufacture, by spot checks in the factory at the Administration's discretion, or for the purpose of accepting a delivery under contract.

They shall include both tests on the materials constituting the bearings and tests on the bearings themselves.

1.5 *Methods of acceptance*

These shall be in accordance with the rulings in force on each Administration. The following procedure can be recommended, however.

1.5.1 Checks by the supplier

For each type of bearing accepted, the supplier shall undertake to advise the Administration (or the inspecting authority appointed by it) of the results of all the tests which he has carried out during manufacture.

The frequency of such checks by the factory shall be at least one full series of tests as described in Section 3 below, on three samples taken from each batch of 150 bearings.

The Administration's inspectors may be present at the tests carried out by the supplier. They may verify that a manufactured batch selected by them conforms to the values guaranteed by the supplier for the authorized type of bearing.

1.5.2 Acceptance of a small quantity

A small delivery is one for an order not exceeding twenty bearings. In this case, a very simple acceptance test may be adopted, consisting of submitting all the bearings in the order to a compression test in accordance with the conditions of Article 3.3.3 below, and half the bearings to a shear test in accordance with the conditions of Article 3.3.4 below.

1.5.3 Acceptance of a large quantity

For a large delivery (more than 20 bearings), the order must stipulate a certain number of additional bearings to enable the Administration's receiving inspector to carry out the full series of tests described in Section 3 below on the samples selected by him from the batch.

The number of additional bearings shall be:

three for an order of 20 to 100 bearings;

six for an order of more than 100 bearings.

The results of these tests shall be recorded in the internal Inspection Register kept by the manufacturer, together with the date of the order and the name of the Administration carrying out the acceptance.*

* To take particular account of cases where the supplier is authorized by several Administrations.

2 Definition of the materials

2.1 *Rubber*

2.1.1 Type

Natural or synthetic rubber may be used. It must satisfy any local regulations and standards in force in each country.

2.1.2 Special properties

In particular, the rubber must have a good resistance to the action of oils, weather, atmospheric ozone and extreme temperatures to which the bearing is subjected.

Natural rubbers are usually able to withstand variations in temperature, particularly low and extreme temperatures (−30°C, −40°C), satisfactorily.

The composition of synthetic rubbers must be designed in relation to the use for which they are intended. The behaviour of the rubber at low temperatures is checked both on the rubber used and on the type of bearing envisaged.

The behaviour of the rubbers at high temperatures (+50°C) calls for no special remarks. For structures situated in certain areas, resistance to certain biological agents (insects, cryptogams, etc.) may be the subject of special specifications.

2.1.3 Mechanical properties

The rubber is characterized from the mechanical point of view by its shear modulus measured in accordance with the conditions laid down in 3.3.4 below.

The G modulus and the Shore A hardness of the rubber, often used for acceptance tests, usually vary in the same direction. The following comparison (at a temperature of 20°C) is given as a guide:

Shore A hardness	50	60	70
G modulus (in bars)	5	88	11

For bearings for railway bridges, the Shore A hardness is generally 60 ± 5.*

* It may be accepted that the readings given by the Shore A durometer are approximately the same as the readings in international degrees of rubber hardness (IRHD) to which certain standards refer.

2.2 *Steels for plates*

2.2.1 Mechanical properties

The steel used for the plates must have:

a yield stress of not less than 2400 kgf/cm^2;

an ultimate strength of not less than 4200 kgf/cm^2.

2.2.2 Thickness of steel plates

The steel plates used as reinforcement must be at least one millimetre thick.

2.2.3 Machining of steel plates

The edges of the plates and any holes provided in them shall be carefully machined to avoid any notch effect.

2.3 *Adhesives*

Any adhesives used to bond together steel plates and the rubber pads or, where applicable, two steel plates, must possess at least the same shear strength as the rubber used. They must also be fully resistant to the action of oils, weather, atmospheric ozone, biological agents and extreme temperatures to which the bearing may be subjected.

Generally speaking, adhesives used to secure steel to steel are high-viscosity resins which set cold and are solvent-free.

2.4 *Other materials*

The use of materials other than those specified above must be subject to prior agreement by the Administration.

3 Acceptance tests

3.1 *Selection and preparation of test pieces*

Bearings intended for testing must be large enough to permit selection of the various test pieces described in 3.2.2 and 3.3 below.

Before testing, the parts selected must undergo artificial ageing in an oven for seven days at a temperature of 70°C ± 1 degC.

3.2 *Tests on materials*

3.2.1 Purpose of the tests

The properties of the test pieces intended for the tests on materials, together with the procedure for carrying out each of these tests, shall be those given in the norms in force on each Administration.

3.2.2 Tests on the rubber (natural or synthetic)

3.2.2.1 Nature of the tests

The tests on the rubber shall include, in addition to the hardness, a stress–strain curve with measurement of the ultimate strength and elongation at rupture and, where applicable, verification that the quality remains the same at low temperatures. The results of these various tests must be at least equal to the values laid down by the Administration at the time of authorization.

3.2.2.2 Selection of test pieces

The test pieces required for the following tests shall be selected from the centre layer of one of the bearings making up the selection referred to in 3.1. Selection shall be carried out by sawing parallel to the metal reinforcement. The thickness of the small pad obtained must not be more than 2 millimetres less than the thickness of the rubber between the two metal plates on either side of the central layer. The surfaces of the small pad obtained shall be carefully ground after sawing and rendered smooth and parallel, taking care to avoid any heating during this operation.

The plan dimensions of the small plate must be large enough to allow five standard tensile test pieces to be cut with a hollow punch.

3.2.2.3 Hardness

The hardness is determined in accordance with the standards. The hardness of the rubber must not differ by more than 5 international degrees of IRHD hardness from the hardness laid down when authority was given.

3.2.2.4 Stress–strain curve

The tensile stress–strain curve with measurement of the ultimate strength and elongation at rupture is effected in accordance with the tensile test standards.

The ultimate strength and elongation at rupture are obtained by taking the average of test results on five tensile test pieces (after ageing). These averages must show the minimum values fixed by the Administration for the type of bearing in question.

3.2.2.5 Behaviour at low temperatures

In abeyance.*

* At the present time, cold tests on rubber are laboratory tests requiring test pieces prepared with great accuracy. Such tests are not practicable, in view of their difficulty, under current acceptance procedures.

3.2.3 Tests on steel plates

These tests consist of determining the tensile curve with measurement of the yield stress and the ultimate strength.

The steel must have:

a yield stress of not less than 2400 kgf/cm^2;

an ultimate strength of not less than 4200 kgf/cm^2.

3.3 *Tests on bearings*

3.3.1 Nature of the tests

The tests on bearings shall include:

determination of the vertical stress–strain curve;

determination of the modulus G;

determination of the ultimate strength in compression;

verification of the bond between the rubber and the metal plates;

where applicable, verification of low-temperature behaviour.

3.3.2 Selection of test pieces

The test pieces required for the following tests shall be selected from the bearing(s) making up the selection referred to in 3.1. These test pieces must have the same total thickness as the bearing pads forming the unit to be tested, or must be made up of at least two layers of rubber and three metal plates, the total thickness of rubber being at least 15 mm. The plan dimensions of the test pieces must be at least 150 × 200 mm.

In the text below, these test pieces will be termed bearings.

The bearings must have undergone artificial ageing before testing under the conditions laid down in 2.1.3.

3.3.3 Determination of the vertical stress–strain curve

This is carried out simultaneously on two bearings placed between three concrete slabs.

Loading must take place in stages, at a rate not exceeding 2 tonnes/min. The distortion is measured by means of gauges arranged at the four corners of each bearing.

The test must be continued up to the limiting stress $\sigma_{m\,max}$ fixed by the Administration.

The test is then repeated.

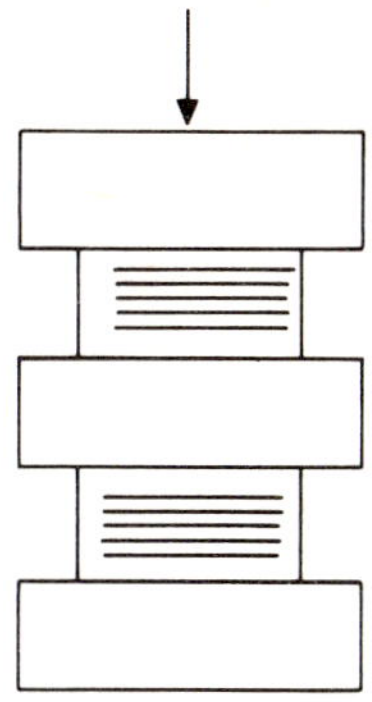

The vertical stress–strain curve is plotted during the second loading. The deflection of the bearing is recorded between the minimum pressure $\sigma_{m\,min}$ corresponding to the loading of the bearings and the load limit $\sigma_{m\,max}$ referred to above. The pressure $\sigma_{m\,min}$ must not be less than 20 kgf/cm^2 (no slip pressure).

The test is considered satisfactory if the values obtained do not differ from the value laid down by more than 20%.

3.3.4 Determination of the modulus G

The modulus G is defined using the equipment and bearings used for the compression test in 3.3.3 above. The upper and lower concrete slabs are securely fixed at the sides to the frame of the press, while a horizontal force is applied to the centre plate.

To prevent slipping of the bearings, a constant vertical load P of 50 kgf/cm^2 is applied.

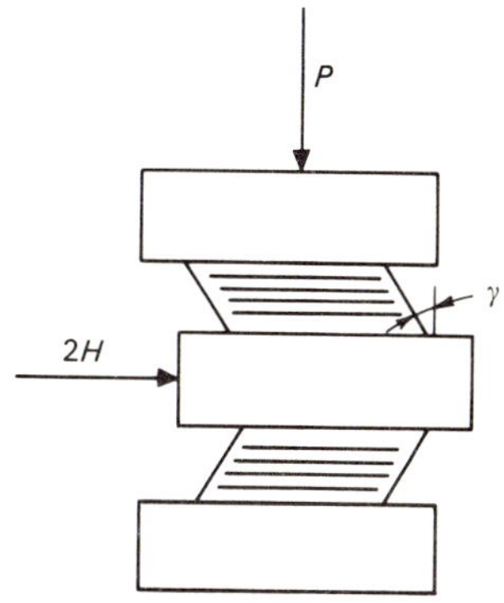

The horizontal force is applied in stages and the horizontal deflection for each stage is measured. The test must be continued until a shear angle γ such that $\tan\gamma = 0{\cdot}9$ is obtained. The rate of loading must not exceed 1 tonne/min.

The test is carried out twice, and the modulus G is determined from the second loading. Since the horizontal stress–strain curve is generally not linear in its lower part, the modulus G is usually determined from the section between $\gamma = 15°$ and $\gamma = 30°$, where γ is the shear angle of the bearing.

The following equation is taken as a basis for calculating the modulus G:

$$G = \frac{\tau}{\tan\gamma}$$

in which

$$\tau = \frac{H}{A} \quad \text{and} \quad \tan\gamma = \frac{u}{T}$$

where

H = horizontal force

A = area of the bearing

u = horizontal deflection

T = height of the actual bearing, i.e. total height of the bearing, less the thickness of the steel plates.

The test shall be considered satisfactory if the values obtained do not differ from the value laid down by more than 15%.

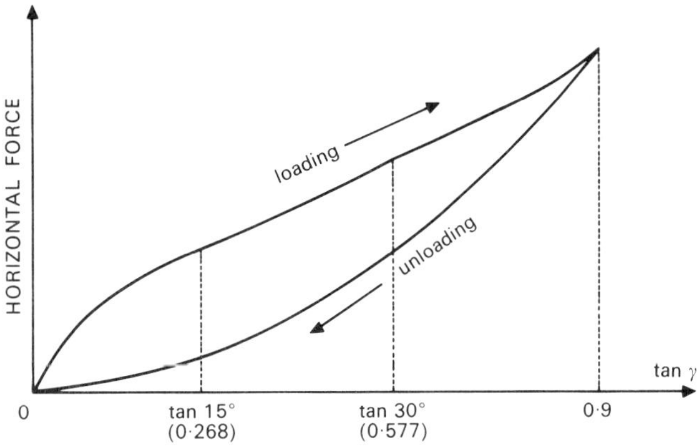

3.3.5 Determination of the ultimate compressive strength
The ultimate strength in compression is determined by applying a compression load in stages to a test piece 100 × 100 mm.

The test piece is placed directly between the plates of the testing machine. The speed of application must not exceed 20 tonnes/min. (Failure generally occurs by fracture of the steel plates, marked by a clearly audible sound.)

The ultimate strength must be not less than six times the working stress σ_m laid down for the bearing.

3.3.6 Test of the bond between the rubber and the steel plates
To check the bond between the rubber and the steel plates, or between the steel plates, following the test described in 3.3.4, the load P is raised to 100 kgf/cm², and the horizontal force increased to obtain a horizontal deflection corresponding to $\tan\gamma = 2{\cdot}0$.

The test must not cause any bond failure.

These tests may be replaced, however, by ultimate load tests on samples of smaller dimensions taking either from the edge or from the centre of the bearing. The ultimate strength must correspond to a shear stress of

$$\tau > \left(1 + \frac{\lambda}{100}\right) G$$

where λ is the percentage elongation at fracture.

Any permanent deformation resulting from this test shall not be considered as a cause for rejection.

3.3.7 Tests to check behaviour at low temperatures
Such tests are in abeyance.

Tests for checking the behaviour of the bearings at low temperatures, such as those effected by the ORE D 60 Committee of Experts, are long and costly and difficult to use for acceptance purposes. However, if the supplier plans to modify the formula of the rubber used for the bearings, he must so advise the Administration. The latter shall then have tests made to ensure that the change in the composition of the rubber does not have an adverse effect on the characteristics fixed when the bearing was authorized.

4 Dimensional tolerances

4.1 Plan dimensions
Tolerances on length and width: −0, +5 mm.

4.2 Thicknesses

4.2.1 Thickness tolerances for a single layer

average value = nominal value ± 0·5 mm

value at any point = average value ± 0·5 mm

These tolerances may be allowed for a single layer but not cumulatively.

4.2.2 Tolerances on the total thickness h of the bearing

$10 < h \leqslant 30$ mm:	±0·5 mm
$30 < h \leqslant 50$ mm:	±0·8 mm
$50 < h \leqslant 80$ mm:	±0·9 mm
$80 < h \leqslant 120$ mm:	±1·1 mm

4.2.3 Tolerances on the thickness t_s of each steel plate
Tolerances are fixed by the corresponding national standard for such steel plates.

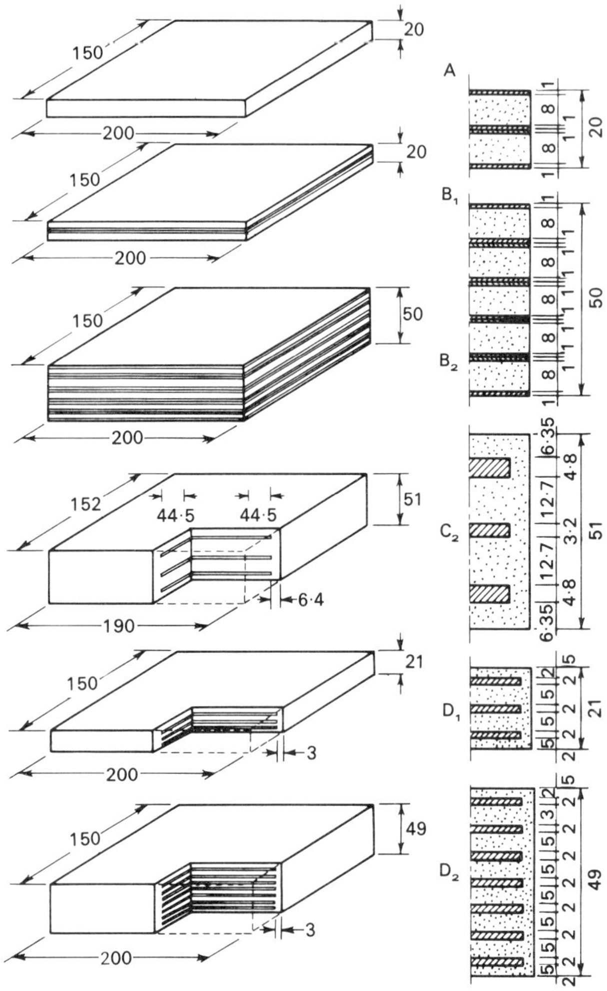

Appendix 2: Design example

A ***Data***
A bearing is to be designed to satisfy the following conditions:

1 Loads and deflections

(*a*) *Slow loads (dead weight, temperature, shrinkage)*

$$P_c = 60{,}000 \text{ daN}$$

$$\tan\alpha_c = -0{\cdot}5 \times 10^{-3}$$

(taking into account manufacturing camber)

$$u_c = 1 \text{ cm}$$

(*b*) *Quickly applied loads (live loads)*

$$P_s = 40{,}000 \text{ daN}$$

$$\tan\alpha_s = +3 \times 10^{-3}$$

$$u_s = 0{\cdot}5 \text{ cm}$$

(extension of the tension flange)

2 Width of the bearing

$b = 52$ cm [cf. clause 3.1]

3 Quality of the rubber [cf. clause 8.13]

$$G = 11 \text{ bars}$$

$$\sigma_{\text{allowed}} = 90 \text{ bars}$$

These two latter values may be changed if the following calculation shows this to be necessary.

B Plan dimensions [cf. clause 3.1]

1 Limitation of the average pressure on the bearing

$$\sigma_m = \frac{P}{A'} = \frac{P_c + P_s}{(a-u)b} < 90 \text{ bars}$$

with

$$u = u_c + u_s = 1 + 0{\cdot}5 = 1{\cdot}5 \text{ cm}$$

whence

$$a - u > \frac{100{,}000}{90 \times 52} \quad \text{or} \quad 21{\cdot}4 \text{ cm}$$

and

$$a_{\min} = 21{\cdot}4 + 1{\cdot}5 = 22{\cdot}9 \text{ cm}$$

We take

$$a = 25 \text{ cm}$$

whence

$$\sigma_m = \frac{100{,}000}{(25 - 1{\cdot}5)52} = 81{\cdot}7 \text{ bars} < 90$$

2 No slip-condition [cf. clause 3.1.2.2.1]

(a) $$\sigma'_m = \frac{P_c}{A'} = \frac{P_c}{(a-u_c)b} > 20 \text{ bars under } P_c \text{ only}$$

$$\sigma'_m = \frac{60{,}000}{(25-1)52} = 48{\cdot}1 \text{ bars} > 20 \text{ bars}$$

condition satisfied.

(b) $$H_c < f'P_c \quad \text{where} \quad H_c = GA\frac{u_c}{T}$$

[cf. clause 3.1.2.2.2]

and

$$f' = 0{\cdot}10 + \frac{2}{\sigma'_m} = 0{\cdot}10 + \frac{2A'_c}{P_c}$$

whence

$$T > \frac{GAu_c}{2A' + 0{\cdot}10P_c}$$

or

$$T > \frac{11 \times 52 \times 25 \times 1}{[2 \times (25-1) \times 52] + [0{\cdot}10 \times 60{,}000)}$$

$$T > 1{\cdot}68 \text{ cm}$$

(c) $H_c + H_s < f(P_c + P_s)$ [cf. clause 3.1.2.2.2]

where

$$H_c + H_s = \frac{GA(u_c + 2u_s)}{T}$$

and

$$f = 0{\cdot}10 + \frac{2}{\sigma_m} = 0{\cdot}10 + \frac{2A'}{P_c + P_s}$$

whence

$$T > \frac{GA(u_c + 2u_s)}{2A' + 0{\cdot}10(P_c + P_s)}$$

or

$$T > \frac{11 \times 52 \times 25 \times [1 + (2 \times 0{\cdot}5)]}{[2 \times (25 - 1{\cdot}5) \times 52] + [0{\cdot}10 \times (60{,}000 + 40{,}000)]}$$

$$T > 2{\cdot}30 \text{ cm}$$

C Thickness

Limitation of lateral movement (cf. clause 3.2.2.1]

$$\tan \gamma_{c,s} < 0{\cdot}70 \quad \text{or} \quad T > 1{\cdot}43(u_c + u_s)$$

$$T > 1{\cdot}43 \times 1{\cdot}5 \quad \text{or} \quad 2{\cdot}15 \text{ cm}$$

D Type of bearing

Three types of bearing would satisfy the above conditions:

(a) one reinforced pad of 24 mm
(b) two reinforced layers of 12 mm each
(c) three reinforced layers of 8 mm each

The following checks are to be made for the three types (a), (b) and (c).

E Vertical deflection

1 Calculation of S_i

$$S_i = \frac{ab}{2t_i(a+b)} = \frac{25 \times 52}{2t_i(25+52)} = \frac{8{\cdot}44}{t_i}$$ [cf. clause 3.2.2.2]

(a) $t_i = 2{\cdot}4$ $S_i = 3{\cdot}51$
(b) $t_i = 1{\cdot}2$ $S_i = 7{\cdot}03$
(c) $t_i = 0{\cdot}8$ $S_i = 10{\cdot}55$

2 No-uplift condition [cf. clause 3.2.2.2]

This condition is first checked for the most rigid type of bearing: Three layers of 8 mm with $t_i = 0{\cdot}8$

(a) $$|\tan \alpha_c| < \frac{6\Sigma e'_i}{a}$$ [cf. clause 3.2.2.2.1]

or

$$\Sigma e'_i > \frac{0{\cdot}5 \times 10^{-3} \times 25}{6}$$

whence

$$\Sigma e'_i > 0{\cdot}0021$$

with

$$e'_i = \frac{t_i\sigma'_m}{4GS_i^2 + 3\sigma'_m}$$

$$= \frac{t_i}{(4G/\sigma'_m)S_i^2 + 3} = \frac{t_i}{(44/48{\cdot}1)S_i^2 + 3}$$

$$t_i = 0{\cdot}8 \qquad S_i = 10{\cdot}55$$

whence

$$e'_i = \frac{0{\cdot}8}{0{\cdot}915(10{\cdot}55)^2 + 3}$$

$$= 0{\cdot}0076$$

and

$$\Sigma e'_i = 3 \times 0{\cdot}0076 = 0{\cdot}028 > 0{\cdot}0021$$

(a) $$\tan \alpha_c + 1{\cdot}5 \tan \alpha_s < \frac{6\Sigma e_i}{a}$$ [cf. clause 3.2.2.2.2]

or

$$\Sigma e_i > \frac{(-0{\cdot}5 + 3 \times 1{\cdot}5)10^{-3} \times 25}{6}$$

whence

$$\Sigma e_i > 0{\cdot}0167$$

with

$$e_i = \frac{t_i\sigma_m}{4GS_i^2 + 3\sigma_m} = \frac{t_i}{(4G/\sigma_m)S_i^2 + 3}$$

$$= \frac{t_i}{(44/81{\cdot}7)S_i^2 + 3}$$

whence

$$e_i = \frac{0{\cdot}8}{0{\cdot}538(10{\cdot}55)^2 + 3} = 0{\cdot}0127$$

and

$$\Sigma e_i = 3 \times 0{\cdot}0127 = 0{\cdot}0381 > 0{\cdot}0167$$

The condition is satisfied for the bearing of type (c).

The less rigid types of bearings (a) and (b) will satisfy this condition even more easily.

3 Stability condition

$$\sigma_m < \frac{2u}{3T} G S_i \quad \text{where} \quad S_i > \frac{3T\sigma_m}{2aG} \qquad \text{[cf. clause 3.2.2.3]}$$

$$S_i > \frac{3 \times 2{\cdot}4 \times 81{\cdot}7}{2 \times 25 \times 11}$$

whence $S_i > 1{\cdot}07$.

Condition satisfied for the three types of bearing (a), (b) and (c) referred to in Section D.

4 Limitation of the shear stress in the rubber

$$\tau = \tau_p + \tau_H + \tau_\alpha \leqslant 5G \quad \text{or} \quad 55 \qquad \text{[cf. clause 3.2.2.4]}$$

with

$$\tau_p = \frac{1{\cdot}5\,P_c + 1{\cdot}5\,P_s}{S_i \quad ab}$$

$$= \frac{1{\cdot}5 \times 120{,}000}{25 \times 52\,S_i} = \frac{138{\cdot}5}{S_i}$$

$$\tau_H = G\frac{u}{T} = \frac{11 \times 1{\cdot}5}{2{\cdot}4} = 6{\cdot}86$$

$$\tau_\alpha = \frac{Ga^2}{2t_i T}(\tan\alpha_c + 1{\cdot}5\tan\alpha_s)$$

$$= \frac{11(25)^2}{4{\cdot}8\,t_i}[-0{\cdot}5 + (1{\cdot}5 \times 3)]10^{-3}$$

$$= \frac{5{\cdot}73}{t_i}$$

(a) $t_i = 2{\cdot}4$ $\quad \tau = 39{\cdot}46 + 6{\cdot}86 + 2{\cdot}39 = 48{\cdot}71$

(b) $t_i = 1{\cdot}2$ $\quad \tau = 19{\cdot}70 + 6{\cdot}86 + 4{\cdot}77 = 31{\cdot}33$

(c) $t_i = 0{\cdot}8$ $\quad \tau = 13{\cdot}13 + 6{\cdot}86 + 7{\cdot}17 = 27{\cdot}16$

The condition $\tau \leqslant 55$ is therefore satisfied for the three types of bearing (a), (b) and (c).*

* In certain cases, it may also be necessary to check the shear stress when the slow load acts alone.

F Choice of solution

Since all three types of bearing have satisfied the checks in Section E, the type (c) bearing may be chosen, for example, comprising three layers of 8 mm.

In actual fact, this type gives the greatest safety margin with regard to uplift, instability and shear.

G Design of reinforcement

$$t_s > \frac{2(t_1 + t_2)(P_c + 1{\cdot}5\,P_s)}{A'\sigma'_a} \qquad \text{[cf. clause 3.2.2.5]}$$

with

$$\sigma'_a = 1800 \text{ bars}$$

1 Intermediate reinforcement

$$t_s > \frac{2 \times 1{\cdot}6 \times (60{,}000 + 1{\cdot}5 \times 40{,}000)}{(25 - 1{\cdot}5) \times 52 \times 1800}$$

$t_s > 0{\cdot}175$ cm or $t_s = 2$ mm

2 Top and bottom reinforcement

$$t_s > \frac{2 \times 0{\cdot}8 \times (60{,}000 + 1{\cdot}5 \times 40{,}000)}{(25 - 1{\cdot}5) \times 52 \times 1800}$$

$t_s > 0{\cdot}087$ cm or $t_s = 1$ mm

Appendix 3: Notation and symbols

1 Material

'Static' shear modulus	G
'Dynamic' shear modulus	G_s
Modulus of elasticity of rubber in laterally unrestrained compression	E
Compression modulus of complete bearing	E_a

2 Geometrical characteristics

Number of layers of rubber	n
Number of layers of rubber, if layers of different thicknesses are used in the same bearing	$n_1, n_2, \ldots$
Dimension of bearing measured in the direction of span	a
Dimension of bearing measured at right-angles to the span	b
Thickness of an individual layer of rubber	t_i
Thickness, if layers of different thickness are used in the same bearing	$t_1, t_2, \ldots$
Total thickness of rubber in bearing	$T = \Sigma t_i$
Total thickness of the bearing	h
Cross-sectional area of the bearing	A
Reduced cross-sectional area of the bearing	$A' = (a - u)b$
Shape factor of the bearing	S
Shape factor of an individual layer of rubber	S_i
Dimensions of steel plates:	
Measured in the direction of the span	a_s
Measured at right-angles to the span	b_s
Thickness	t_s
Plate area	A_s

3 Loads and moments

Vertical load	P
Slow vertical load	P_c
Quick acting vertical load	P_s
Horizontal load	H
Horizontal slow load	H_c
Horizontal quick acting load	H_s
Moment, caused by rotation of bearing	M
Coefficient of friction	f or f'

4 Stresses

Compressive stresses	σ
Mean compressive stress under the action of a slow load	σ'_m
Mean compressive stress under the action of slow and quick acting loads	σ_m
Shear stress	τ
Mean shear stress due to a horizontal force H	τ_H
Shear stress due to a vertical force P	τ_p
Shear stress due to a rotation α	τ_α
Permissible tensile stress in the steel	σ'_a

5 Deformations

Compressive strain	$\varepsilon = (\Delta T)/T$
Total deflection under compression under the simultaneous action of slow loads and quick acting loads (unreinforced bearings)	e
Total deformation under compression under the action of a slow load (unreinforced bearings)	e_c
Total deformation under compression under the simultaneous action of slow loads and quick acting loads (reinforced bearings)	e_i

Total deformation under compression under the action of a slow load (reinforced bearings) . . . e_i'
Shear strain (horizontal movement) . . . . tan γ
Shear strain under the simultaneous action of horizontal slow and quick acting forces . . . . . tan $\gamma_{c,s}$
Horizontal shear movement . . . . u
Horizontal shear movement due to horizontal slow acting forces . . . . u_c
Horizontal shear movement due to a quick acting horizontal force . . . . u_s
Horizontal shear movement of the various rubber layers of different thicknesses . . . . $u_1, u_2, \ldots$
Rotation . . . . α
Rotation under the effect of slow loads . . . . α_c
Rotation under the effect of quick acting loads. α_s
Rotation of a layer . . . . $\alpha_1, \alpha_2, \ldots$
Ultimate elongation of the rubber . . . . λ

6 *Special notation*
The suffixes 'min' or 'max' will be used to designate the minimum or maximum values . . . . . P_{max}, σ_{min}
The suffix 'adm' will be added to the symbols to indicate permissible values. . . . . α_{adm}
The suffix '*c*' added to a load or deformation symbol relates to slowly changing loads . . . . P_c, u_c
The suffix '*s*' added to the same symbols relates to quickly applied loads . . . . P_s, u_s

Application
All Administrations in the Union.